Deregulating Telecommunications

CRITICAL MEDIA STUDIES
INSTITUTIONS, POLITICS, AND CULTURE

Series Editor
Andrew Calabrese, University of Colorado at Boulder

Advisory Board
Patricia Aufderheide, American University • **Jean-Claude Burgelman**, Free University of Brussels • **Simone Chambers**, University of Colorado • **Nicholas Garnham**, University of Westminster • **Hanno Hardt**, University of Iowa • **Gay Hawkins**, The University of New South Wales • **Maria Heller**, Eötvös Loránd University • **Robert Horwitz**, University of California at San Diego • **Douglas Kellner**, University of California at Los Angeles • **Gary Marx**, M.I.T. • **Toby Miller**, New York University • **Vincent Mosco**, Carleton University • **Janice Peck**, University of Colorado • **Manjunath Pendakur**, University of Western Ontario • **Arvind Rajagopal**, New York University • **Kevin Robins**, Goldsmiths College • **Saskia Sassen**, University of Chicago • **Colin Sparks**, University of Westminster • **Slavko Splichal**, University of Ljubljana • **Thomas Streeter**, University of Vermont • **Liesbet van Zoonen**, University of Amsterdam • **Janet Wasko**, University of Oregon

Deregulating Telecommunications

U.S. and Canadian Telecommunications,
1840–1997

Kevin G. Wilson

ROWMAN & LITTLEFIELD PUBLISHERS, INC.
Lanham • Boulder • New York • Oxford

ROWMAN & LITTLEFIELD PUBLISHERS, INC.

Published in the United States of America
by Rowman & Littlefield Publishers, Inc.
4720 Boston Way, Lanham, Maryland 20706
http://www.rowmanlittlefield.com

12 Hid's Copse Road, Cumnor Hill, Oxford OX2 9JJ, England

British Library Cataloguing in Publication Information Available

Library of Congress Cataloging-in-Publication Data
Wilson, Kevin G., 1952–
    Deregulating telecommunications : U.S. and Canadian
telecommunications, 1840–1997 / Kevin G. Wilson.
      p. cm.—(Critical media studies)
    Includes bibliographical references and index.
    ISBN 0-8476-9824-6 (alk. paper)—ISBN 0-8476-9825-4 (pbk. : alk. paper)
      1. Telecommunication—Deregulation—United States. 2.
  Telecommunication—Deregulation—Canada. I. Title. II. Series.

HE7645 .W55 2000
384'.0973—dc21

                              00-038269

Printed in the United States of America

∞™ The paper used in this publication meets the minimum requirements of American National Standard for Information Sciences—Permanence of Paper for Printed Library Materials, ANSI/NISO Z39.48-1992.

To my parents Joan and Gerry Wilson who witnessed and were participants in many of the events described in this book as employees of the New Jersey Bell Telephone Company from 1935 to 1988.

Contents

Contents

**Part II: From Monopoly to Competition: The Deregulation of U.S.
and Canadian Telecommunications, 1946–1997**

Figures and Tables

FIGURES

TABLES

Preface

Writing in the *Canadian Journal of Communications* in 1985, Vincent Mosco noted, "the challenge to teachers of telecommunications policy is to define a new field that fits no well established discipline" (51). Over ten years later, the challenge is still formidable. The field of telecommunications policy has traditionally been the privileged domain of the disciplines of law and economics. The application of these disciplines to telecommunications policy has not meant that the field of telecommunications policy is inaccessible to students of communications. But it has meant that some familiarity with these disciplines is required if students are to understand and eventually participate in the policy process.

There are two other obstacles. The student of communications must have a basic understanding of some the technical aspects of telecommunications systems. This knowledge need not be extensive or detailed. For the most part, it is simply a case of mastering the vocabulary of telecommunications. But it is a vocabulary that is not generally taught in the communications curriculum. Finally, students of communications are generally not familiar with the history of the telephone industry and its regulation.

The history of the mass media and of public policy for the mass media are generally part of the communications curriculum at the graduate level; to a lesser extent they may also be taught as part of the undergradute curriculum at the advanced level. But the history of the telephone industry, its regulation, and a century-old public policy concerning the industry are not part of the basic communications curriculum. This book is designed to fill this void in communications studies. It does this by analysing the history of the transition from monopoly to competition in the Canadian and U.S. telecommunications industries using a critical approach.

Why is it important for students of communications to be familiar with the history of the telecommunications industry? The answer in a single word is *networks*. The evolution of modern communications systems is inextricably linked to the problem of network development and, in particular, to the development of high-capacity, broadband networks capable of transmitting voice, images, and high-speed data. Who will build the networks? Who will maintain them? Who will control them? Who will have access to them and at what price? The answers to these questions are not to be found in the characteristics of the technology. Although the discourse of technological determinism encourages us to believe that technology is the driving force behind the emergence of new communications and information services, this is only half of story.

The history of the telecommunications industry, of its regulation and deregulation, suggests that the transformation of technology into networks (facilities) and services takes place in a vigorously contested arena of government regulation and other institutions of public control. Private sector interests compete for economic advantage before regulatory boards, government agencies, and the courts. The rules of this game, played out before public sector institutions, are established in laws, by the courts, by the rules of regulatory procedure and due process, as well as through the competition of ideas crafted by policy institutes, such as the American Enterprise Institute, whose activities are funded by powerful economic interests.

Private sector investment in technology is transformed into networks and services through a complex process of rule making and arbitration involving the legislative, judicial, and regulatory institutions of the state. These processes determine the structural organization of the supply of telecommunications facilities and services, most notably, the extent of horizontal and vertical integration (concentration) of the industry. This in turn determines the price of services and, ultimately, whether they will be available to different groups in society. A critical understanding of contemporary telecommunications requires an understanding of these public sector institutions, of their relationship to private sector actors, and of the processes and policies that lead to decisions concerning such matters as rates for services, subsidies, the licensing of the radio spectrum, and the rules that govern the interconnection of networks, to name a few.

It is a basic premise of our approach that the students of telecommunications need to understand the policy-making process, and, in particular, how the courts and the regulatory agencies have shaped the transformation of the industry from monopoly to competition through a cumulative process of case-by-case decision making. Our emphasis on the process of change has meant that we have been careful not to glance over, or omit, basic information that will enable the reader to situate events, to place them in context, and to relate them to one another.

Accordingly, we have been careful to provide specific dates for events, such as the issuing of regulatory decisions. Where appropriate we have provided information about groups and individuals who have intervened in regulatory proceedings or otherwise played an important role in shaping policy. We have been careful to avoid the gratuitous use of technical jargon. Where technical terms are employed we have taken care to define them in layman's language. Finally, we have attempted, wherever possible, to document our research using primary sources. This has been done to facilitate historical research into telecommunications by researchers and students of communications who will come after us, and may wish to use our work as a basis for further historical analysis. This attention to detail is not meant to overburden the reader with factual information, rather it is designed to provide the reader with a relatively complete picture of the processes (regulatory, judicial, and legislative) through which structural change has been introduced to the telecommunications sector.

Acknowledgments

I would like to acknowledge the valuable contribution that Robert E. Babe's work has made to our work. An economist and professor of communications at the University of Windsor (Ontario), Babe's studies into the history of Canadian telecommunications have laid a crucial foundation for critical studies on Canadian telecommunications.

Pablo Cantera of Montreal assisted in preproduction copyediting and produced the index. It was a pleasure to be seconded by such a conscientious and professional individual at these crucial stages of the book's production.

This book was made possible by a grant from the Social Sciences and Humanities Research Council of Canada. We gratefully acknowledge their support.

General Introduction

During the last week of November 1999, the World Trade Organization (WTO) held its Third Ministerial Conference in Seattle, Washington. The William J. Clinton administration saw the meeting as a tailor-made opportunity to showcase its policy of freer trade among nations. Instead, the gathering produced a riot. The conference became a magnet for citizen groups from around the world, who gathered in the streets outside the conference to protest the elitist and undemocratic nature of the conference.

Social discourse on globalization is heavily laden with determinist suppositions. In their *Cyberspace and the American Dream: A Magna Carta for the Knowledge Age*, Esther Dyson, George Gilder, George Keyworth, and Alvin Toffler herald the arrival of the Third Wave economy (Dyson 1994). The authors contend that new information technologies are the driving force behind a post Second Wave, postindustrial, knowledge economy, which is leading to a demassification of institutions and culture. According to these authors, the emergence of cyberspace is compelling society to reexamine: the meaning of freedom, structures of self-government, the definition of property, the nature of competition, the conditions for cooperation, the sense of community, and the nature of progress. While these are legitimate preoccupations, Third Wave social theory frames them in terms of a theory of technology-society relations that is essentially historicist (Kumar 1995) and determinist. This approach would have us believe that technology is propelling society to the next stage of social evolution. Third Wave theorists are merely sounding the alarm. Society must prepare. It must adapt. Through an analysis and comparison of U.S. and Canadian telecommunications deregulation, this study brings to light a more complex dynamic in which economic, political, and social interests clash and compete in an ongoing, incremental process of policy making and technology implementation.

Fifty years after the invention of the first electronic computer, we have entered the era of digital communication. The explosive growth of the Internet is probably the most potent symbol of the new mode of commerce, exchange, and communication. Advances in information and communications technologies have provided the infrastructure for the new global economy, but they have not propelled the economy or society on a predetermined course. Public sector institutions have played a key role in determining the pace and the direction of innovation and diffusion of communications technologies since the invention of the telegraph. These institutions are not autonomous; nor do they always act in harmony with one another. In fact, they often work at cross-purposes. They may be subject to political pressure and are often the object of intense lobbying by special interests. Most important, they function within a political economy of state-industry relations circumscribed by constitutional frameworks, regulatory and legislative traditions, legal jurisprudence, and dominant ideology. These institutions operate at the national, state, and to a lesser extent, local levels. Public sector investment and regulation have played a vital role in determining which technologies are developed, and in shaping the industry structures (monopolistic, oligopolistic, and competitive) through which infrastructures are developed and services are brought to market. Historicist and determinist approaches to technology-society relations, such as Third Wave social theory, obscure the ways that these institutions have determined and will continue to determine the price, variety, quality, and accessibility of communication services. Nowhere is this truer than the communications sector encompassing both broadcasting and telecommunications. Moreover, policy making, particularly when it concerns communications, has been a quintessentially national endeavour.

It is difficult to imagine two nations whose economies and cultures are more connected than those of Canada and the United States. Although Canadians are well aware of this relationship, Americans, generally, are not. Moreover, most Canadians are reconciled to the fact that Americans know very little about their country. For example, most Americans do not know that Canada is their principal trading partner. They have never heard of the 1965 Autopact, which has been a centerpiece of Canadian industrial policy since 1965. (The Autopact exempts the U.S. Big Three automobile manufacturers from paying Canada's 6.7 percent import tariff if they engage in local procurement and production.) Nor are most Americans aware that Canada has two official languages.

For Canadians a study that compares and contrasts U.S. and Canadian deregulation of the telecommunications industry has an obvious resonance. U.S. deregulation preceded Canadian deregulation and created pressure to pursue a similar course north of the border. But the impact of U.S. deregulation on the Canadian telecommunications industry was neither immediate nor direct. Canadian deregulation was not a carbon copy of the U.S. experience. From a Canadian perspective, a comparative study brings to light the numerous ways that Canadian policy and institutions have shaped the pace and the course of deregulation. By bringing these determinants into sharper focus, such an analysis heightens our understanding of Canadian traditions, culture, and institutions.

Although U.S. deregulation has influenced policy making in Canada, one would be hard-pressed to argue that Canadian policy has exerted a reciprocal influence on its neighbor. How are we to understand the value of a comparative study from an American perspective? There is the obvious, intrinsic value of enhancing understanding of one's political culture and institutions through the method of comparison and juxtaposition. Globalization and its discontents provide the other context.

The WTO Trade Ministers Conference, and the social protest it elicited, compels an analysis and understanding of the national dimension of communications policy: how national (and to a lesser extent local) interests, myths, traditions, aspirations, and goals are advanced within, and ultimately shaped by the policy process. Globalization seeks to circumvent and deny these forces. The history of deregulation in Canada is a relevant and accessible case study of an alternative model of policy making and policy implementation at the dawn of the digital age. As such it is a potent reminder that a policy for communications need not ignore the needs and aspirations of local, regional, or national communities.

By contrasting Canadian and U.S. deregulation we will cast a critical light on the policy-making institutions of both countries. What is unique about policy-making institutions in each country? What interests have shaped changes in policy and the policy process? And most importantly, what are the specific challenges that policy makers in each country will face in the future?

Part I

FROM COMPETITION TO MONOPOLY: THE
CONSOLIDATION AND REGULATION OF U.S. AND
CANADIAN TELECOMMUNICATIONS, 1840–1946

Introduction to Part I

Today thousands of companies provide telecommunications services throughout the United States and Canada. These services include: public voice service (local and long distance telephony); public mobile (cellular) telephony; public data services (Internet access, America On Line, and so on); and private data and voice services for corporations, to name a few. Many new services such as Asychronous Digital Subscriber Lines (ADSL), Integrated Services Digital Networks (ISDN), cable Internet access, and Personal Communication Services (PCS) have only recently begun to reach the market. Data services (based on digital communications) and mobile telephony are relatively recent additions to the telecommunications landscape. For over a century, from the introduction of commercial telegraph in the late 1840s to the introduction of the first public data services in the mid 1960s, the telecommunications industry was comprised of two industries: telegraph and telephone. In chapters one and two we will consider how the telegraph and telephone industries developed in the United States and Canada. In chapter three we will consider their regulation as public utilities.

The early history of the telegraph and telephone industries is extremely relevant to an understanding of contemporary telecommunications. First and foremost, a review of this period is an excellent introduction to the current industry structure, since many of the firms that established themselves during this period continue to be the dominant players in the industry today. It is important to understand how firms such as AT&T, and the other Bell System companies in the case of the United States, and Bell Canada rose to positions of dominance in their respective national telecommunications industries. Second, the history of this period illustrates the profound impact that U.S. telecommunications has had on the development of telecommunications in Canada. This theme will return in subsequent chapters when we consider the forces that have impelled Canadian deregulation. Third, as we shall

7

see in chapter seven, section two, the profusion of provincial telecommunications monopolies led to problems in the implementation of a national telecommunications policy for Canada. Chapter two provides important background information on the historical factors that divided authority to regulate telecommunications between the provinces and the federal government, and led to calls for a national telecommunications policy for Canada.

Finally, the history of telecommunications during the first one hundred years is the story of the emergence of geographic telephone monopolies. The emergence of these monopolies prompted the regulation of the industry. In order to evaluate conflicting claims concerning the status of telecommunications as a natural monopoly—a debate that has been at the heart of attempts to promote or to oppose the deregulation of telecommunications—it is essential that one understand how these monopolies came into existence in the United States and Canada and why they were subject to economic regulation. In chapter three we will examine the historical justifications for the economic regulation of common carriers and the institutional origins of the agencies that were created to regulate the telephone and telegraph industries.

1

Telegraphs and Telephones: The Birth and Consolidation of the U.S. Telecommunications Industry, 1840–1936

1.1 INTRODUCTION

In this chapter we will recount the early history of the telegraph and telephone industries in the United States. The story begins for each industry with the singular contribution of a gifted inventor: Samuel F. Morse in the case of the telegraph, and Alexander Graham Bell in the case of the telephone. Although innovations in technology were key to the spread of telegraphy and telephony, the actual commercialization of these systems brought into play numerous factors unrelated to the excellence or the efficiency of technology. For example, in order to benefit financially from their work, Morse and Bell obtained patents on their inventions and defended these patents from infringements by competitors. These were legal protections unrelated to the forces of the market. But they would have a profound effect on the structure of the telegraph and telephone industries during the early decades of their commercialization.

The telegraph and the telephone were not off-the-shelf consumer appliances. Their operation and use required an extensive and highly coordinated network of facilities. In this respect, the telegraph and the telephone were akin to other "network" industries such as the railroad, water, gas, and electric industries. These industries share a common trait: they all entail very high initial costs related to investment in essential infrastructure. As a result, the commercial expansion of these industries will depend heavily on the ability to secure financing through partnerships, mergers, alliances, and privileged relations with bankers. The need to increase capitalization becomes a key factor shaping the structure of these network industries. Moreover, as capital requirements increase, the demands of investors for stability and the elimination of risk also increase. The obvious solution from the producer's point of view is monopoly. Although monopoly may be the perfect solution to the problem of how to secure investment, it will usually result in heightened public scrutiny of the industry and may ultimately lead to direct intervention by the government under antitrust law. The extent to which government intervenes in an industry in order to foster competition will also have an impact on industry structure. Technology, patent law, the problem of capitalization, corporate strategy, and government regulation all played a part in determining the structure of the telegraph and telephone industries in the United States during the early years of their commercialization.

1.2 THE ORIGINS AND CONSOLIDATION OF THE U.S. TELEGRAPH INDUSTRY

In 1840 Samuel F. Morse (1791–1872), an artist by profession, received a patent from the U.S. Patent Office for the magnetic telegraph. (The 1840 patent application would be followed by others, the last of which expired by 1861.) Morse had two partners: Professor Leonard Gale, a professor of chemistry at the University of the City of New York, and Alfred Vail of Morristown, New Jersey. Vail provided the capital for the project and assisted Morse in perfecting the telegraph instrument. Gale provided scientific expertise. In 1843 Morse received funds from the U.S. Congress for a demonstration. Telegraph lines were strung between Baltimore and Washington, D.C., and on May 24, 1844, Morse sent the now famous words, "What hath God wrought!" Morse and his partners were unsuccessful in procuring more funding from Congress in spite of the fact that the postmaster general recommended nationalization (Brock 1981, 63). In 1846 Morse and his partners formed the Magnetic Telegraph Company. In the spring of 1846 the company established a commercial telegraph service between Washington, D.C., and New York City.

Morse and his partners exploited the patent through the operations of the Magnetic Telegraph Company, but they also licensed the patent for use by other local telegraph companies in other regions of the country. Until the expiration of the Morse patents, patent rights played a major part in determining industry structure. The genius of the Morse system was its simplicity and reliability: the system worked even when the wires were of inferior quality. The Morse system consisted of a power source (a battery), a relay circuit, a key device for opening and closing the circuit, and a register that produced marks on a piece of paper. A key component of the system was the code developed by Morse to convert the letters of the alphabet into dots and dashes (short and long electrical signals). Competing patents were developed by Royal E. House of Vermont in 1846 and by Alexander Bain of Scotland in 1848. In 1851 the industry was competitive with over fifty different telegraph companies operating in the United States (Harding 1990, 3). But the industry would soon undergo a period of consolidation.

The corporate predecessor to the Western Union Telegraph Company, the New York and Mississippi Valley Printing Telegraph Company (NYMVPTC), was established in 1851. The company was organized by Hiram Sibley of

Rochester New York and Judge Samuel L. Selden, who held the rights to exploit the telegraph patent developed by Royal E. House. The House patent was more complex and less reliable than the Morse system. Unlike the Morse system, the register developed by House printed the letters of the alphabet directly onto a tape. Rather than compete in the eastern corridor of the United States, Sibley and Selden decided to expand operations through the purchase of rival companies operating west of Buffalo. The NYMVPTC operating companies used the House patent. In 1851 it was determined that the Bain patent infringed upon the basic Morse patent, which resulted in a consolidation of the Bain and Morse companies (Brock 1981, 74). In 1855, NYMVPTC succeeded in purchasing lines controlled by Ezra Cornell in a hostile takeover. These lines served the Great Lakes region using the Morse system (77), giving NYMVPTC access to the Morse system. In 1856 the company changed its name to the Western Union Telegraph Company, signifying its western ambitions. In 1855 the American Telegraph Company was formed to compete with Western Union in the east (79). The period 1857 to 1861 was marked by consolidation. By 1861 there were six telegraph groups:

> the American Telegraph Company (covering the Atlantic and some Gulf states), The Western Union Telegraph Company (covering states north of the Ohio River and parts of Iowa, Kansas, Missouri, and Minnesota), the New York Albany and Buffalo Electro-Magnetic telegraph Company (covering New York State), the Atlantic and Ohio Telegraph Company (covering Pennsylvania), the Illinois & Mississippi Telegraph Company (covering sections of Missouri, Iowa, and Illinois), and the New Orleans & Ohio Telegraph Company (covering the southern Mississippi Valley and the Southwest). (Harding 1990, 4)

For the most part, these companies tried to avoid direct competition with one another (3).

In 1860 Congress passed the Pacific Telegraph Act, which authorized the secretary of the treasury to solicit bids for the construction of a transcontinental telegraph line. A subsidy would be provided to the winning bidder. The members of the telegraph cartel, including Western Union and the American Telegraph Company, decided not to enter a joint bid in the hope that the government would increase the amount of the subsidy. However, Western Union broke ranks with the cartel and was awarded the contract as the only bidder. The transcontinental line was completed by Western Union in 1861. Also in that year, the last of the Morse patents expired. This resulted in a further consolidation of the industry. By the end of the Civil War, three companies dominated the industry: Western Union, American Telegraph, and United States Telegraph. In 1866 American Telegraph and United States Telegraph were absorbed into Western Union (Herring and Gross 1936, 2).

The next stage in telegraph expansion coincided with the expansion of the railroad. Initially, the railroads were skeptical of the usefulness of the telegraph.

During the 1850s railroads began to use telegraph lines strung along their rights-of-way to dispatch trains more efficiently. By the 1860s a model had begun to emerge that saw the railroad station agent serve as the commercial representative for the telegraph company. This development was advantageous for the railroads because priority would be given to railroad business. Revenues derived from public telegraph operations were divided between the railroad company and the telegraph company. The telegraph companies were responsible for constructing and maintaining the infrastructure of poles, wires, and station equipment. A key to Western Union's success was the signing of exclusive contracts with railroad companies. It was also able to negotiate an exclusive contract with Associated Press. By 1870 Western Union was distributing almost all the news published in the United States (Herring and Gross 1936, 3).

In 1874 the Atlantic and Pacific Telegraph Company began to challenge the hegemony of Western Union. Atlantic and Pacific was organized in 1867 and began operations with a line connecting New York City to Buffalo. In 1869 Atlantic and Pacific reached an agreement with the Union Pacific Railroad Company. In exchange for stock in the telegraph company, Union Pacific turned over its telegraph operations to Atlantic and Pacific and the latter agreed to extend its lines west to San Francisco (U.S. Supreme Court 1895). By 1874 its operations had reached Omaha, Nebraska. One of the companies Atlantic and Pacific purchased in its western expansion was the Ogden–San Francisco line owned by the Central Pacific Railroad. This company was controlled by the robber baron Jay Gould. Gould wrested control of the Atlantic and Pacific Telegraph Company and then initiated a price war with Western Union designed to weaken the company (Brock 1981, 85). Western Union refused to lower its rates and, as a result, saw its share of the market fall. In 1881 the two companies were merged, but retained the Western Union name. According to a U.S. Bureau of Census report, at the beginning of the 1880s, there were seventy-seven telegraph companies in operation. Eighteen were owned by railroads. The rest were small companies. However, Western Union carried 92 percent of messages and operated 77 percent of the total miles of telegraph lines (Herring and Gross 1936, 11). After 1881 Western Union would dominate the telegraph industry.

By the 1930s the principal domestic competitor to Western Union was the Postal Telegraph-Cable system. (Cable signified overseas telegraphy using underwater cable.) The Postal Telegraph Company was organized to introduce competition to the overseas telegraph market. The impetus for the company came from James Bennet, the owner of the *New York Herald*. Bennet was the largest user of transatlantic cable at this time, and was dissatisfied with Western Union's monopoly rates. Bennet encouraged John W. Mackay, who made his fortune in the Nevada's silver mines, to form a company to lay transoceanic cable. In 1883 the Commercial Cable Company was incorporated (Herring and Gross 1936, 22). When they discovered that they could not distribute messages within the United States via the telegraph, they were forced to use the postal service, ergo the name Postal Telegraph. Eventually, they began to acquire local tele-graph companies, combining them into a common system (Stone 1974). Postal Telegraph coordinated its operations with the domestic system of the Mackay Radio and Telegraph Company. During the 1920s these two companies became part of the International Telephone and Telegraph system created by Sosthenes and Hernando Behn. Other members of the group included All America Cables, which provided service to South America, and the Commercial Cable system, which offered transatlantic service to Europe.

In 1932 Western Union transmitted an average of 7,542,300 telegraph/cable messages monthly. The Postal Telegraph-Cable system carried less than half of that number: 2,343,775 (Herring and Gross 1936, 15). In 1945 Western Union and Postal Telegraph-Cable merged their operations. Thereafter, Western Union separated its domestic and foreign telegraph operations (Harding 1990, 4).

The following illustrates the variety of services that were provided by the tele-graph companies in the 1930s:

1. Market quotations and reports, baseball and other news furnished by mes-senger, private wire, or ticker
2. Money-order service
3. Messenger service
4. Photogram service, covering the facsimile transmission by wire of pictures and messages
5. Marine service, including reports of sighting and arrival of incoming steamships
6. Air express and freight services
7. Travel-check service
8. The handling of aircraft, bus, and theater tickets (Herring and Gross 1936, 11)

1.3 THE ORIGINS AND CONSOLIDATION OF THE U.S. TELEPHONE INDUSTRY

In 1872 Alexander Graham Bell (1847–1922), a teacher of the deaf with a talent for music and an interest in acoustics, began experiments that would lead to the invention of the telephone. Although Bell's goal was to develop a system for telegraphing speech, his first experiments with electricity were at-tempts to improve the telegraph by making it possible to transmit several messages simultaneously over a single wire. (This technique would later be known as multiplexing.) To assist him in the making of parts and equipment, Bell engaged the help of Thomas A. Watson. In Boston, in June 1875, Bell and Watson successfully transmitted sounds over wire for the first time. In the coming months they would perfect the system. On March 7, 1876, Bell received a basic process patent covering the method he had developed for transmitting speech. Several days later Bell succeeded in transmitting human speech. In 1877 Bell received a patent for the basic receiver device.

Under U.S. patent law the instruments that Bell had perfected were protected from use by competitors for sixteen years. Although there were a number of challenges to the Bell patents, none were successful. The most important challenge to the patent was brought by Western Union when the company tried to enter the telephone business. The Bell interests brought a patent infringement suit against Western Union, which was resolved out of court through a settlement. In November 1879 Western Union promised to withdraw from the telephone business in exchange for a royalty of 20 percent on all telephones used in the United States (Stone 1997, 23).

The settlement meant that the Bell interests could exploit the telephone market virtually unchallenged. By reducing the risk to investors, the patents facilitated the commercialization of Bell's invention. But the challenge was daunting as an entire infrastructure of wires, switching centers, and subscriber equipment had to be created from nothing. Improvements to technology, such as the introduction of two-wire circuits (1883) and the perfection of a method of producing hard-drawn wire for telephone lines (1883) improved the technical and economic viability of the basic system (Herring and Gross 1936, 48). The invention of the loading coil and the repeater produced the first long distance call between New York and Chicago in 1892 (53). Nevertheless, the real challenge of commercialization lay in the realm of corporate strategy, a major component of which was the problem of how to finance expansion.

Initially, expansion was financed through investments made by a small group of partners. Thomas Sanders and Gardiner Hubbard, both of Boston, were the first businessmen to provide funds to Bell in exchange for a share of the patent rights. However, the need for capital quickly exceeded the partners' means. As the enterprise grew, it generated a number of corporate forms, each one dictated by the need to increase the capitalization of the company. In quick succession corporate entities were created and transformed. The precursor companies to the American Telephone and Telegraph Company (AT&T), the future telecommunications monolith, were the following: Bell Patent Association (1875), Bell Telephone Company (Massachusetts association) (1877), New England Telephone Company (1878), Bell Telephone Company (1878), National Bell Telephone Company (1879), American Bell Telephone Company (1880), and American Telephone & Telegraph Company (as a subsidiary of the American Bell Telephone Company) (1885). In the short period between the creation of the Bell Telephone Corporation in 1878 and the creation of the American Bell Telephone Company in 1880, capitalization was increased from $450,000 to $7,350,000 (Stone 1997, 26–27). In addition to stock offerings, expansion was financed through the issuance of bonds and debentures.

Financing expansion was a major challenge to the holders of the Bell patent. Equally important was the problem of how to exploit the patent most effectively. This encompassed a number of problems: how to control future innovations in technology, what markets to exploit (e.g., business vs. residential, local vs. long distance), and where to expand. In 1878 Theodore N. Vail (1845–1920) was hired as the general manager of the Bell Telephone Company. By 1880 Vail had developed a plan to exploit the patent and control the emerging telephone industry. First, a parent corporation, the American Bell Telephone Company, was created to handle the licensing of the Bell patent to local telephone companies. Second, because Vail was one of the first to appreciate the value of long distance telephony, the development of a network of long lines became a priority. Finally, he took steps to acquire a manufacturing company (Federal Communications Commission 1939, 29). Eventually, these measures produced what would later be known as the "Bell System," a vertically integrated corporation with AT&T at its corporate head.

American Bell lacked the financial resources to own and operate local telephone companies while simultaneously developing its long distance network. As a result, the local telephone market was developed by licensing local interests to use the Bell patent. Under the license agreements, American Bell would provide the licensee with equipment manufactured under the Bell patent, while the licensee was responsible for constructing and maintaining the lines. Initially these contracts were for a limited period (between five and ten years). However, by 1880 American Bell was signing permanent contracts in exchange for stock in the local enterprises. This process enabled American Bell to conserve capital for the development of its long lines network, while gaining a measure of control over the local market. In 1882 the Bell interests purchased the Western Electric Company of Chicago, a manufacturer of telephone and telegraph equipment. Finally, in 1899 a corporate restructuring saw AT&T (the subsidiary responsible for long lines) acquire all the assets of its parent company, the American Bell Telephone Company (Herring and Gross 1936, 60). Essentially, this was the final restructuring. Thereafter, AT&T was both the parent company of the Bell System and the operator of the long lines network. As parent company, it had an equity interest in the Bell System's local operating companies throughout the United States, and a controlling 40 percent equity interest in Western Electric, who became the sole supplier of equipment to the Bell System. In 1887 Vail left AT&T to pursue other business interests.

When the Bell patents expired in 1892 and 1893 a new era of telephone expansion began. During the period of patent monopoly, the Bell interests had invested primarily in the urban business markets. The expiration of the Bell patents opened the door to competitors seeking to exploit new markets. When so-called independent telephone companies entered the market, they catered to markets in smaller cities and rural areas that the Bell interests had ignored. Most of this growth took place in the midwestern states of Idaho, Illinois, Indiana,

Iowa, Kansas, Minnesota, and Missouri (Herring and Gross 1936, 62). However, a large number of independents also set up exchanges in Bell territory with the hope of competing on the basis of price and service (61). It was during this period that residential service began to grow. In 1893 there were 266,431 telephone "stations" (subscriber lines) in operation; all of them were owned by the Bell companies. Ten years later the Bell System had burgeoned to 1,684,877 telephones, but the total of all independents had nearly matched this expansion with 1,244,936 telephones (Federal Communications Commission 1939, 129). Richard Gabel has appropriately noted that the expiration of the Bell patents "resulted in the most rapid rate of growth of service in the history of the industry" (1969, 342). Although competition was good for subscribers, it hurt Bell's profitability. In 1895 the average yearly revenue per station for Bell was $88. By 1907 this had fallen to $43 (346).

During the early competitive period, 1894 to 1906, the Bell System responded to competition by accelerating the licensing of local exchanges while continuing to invest in its long lines network. In order to undermine the value of the independents' local exchange service, Bell adopted a policy of refusing them interconnection with its long distance service. The Bell System also refused to sell equipment to the independents. Finally, the company launched a propaganda campaign designed to undermine public confidence in the independents (349–350).

In 1907 Vail returned to AT&T. His return coincided with the ascendancy of the Baker-Morgan financial interest within AT&T. In 1902 George Baker, president of the First National Bank of New York, together with associates who included the firm of J. P. Morgan & Co., acquired fifty thousand shares of AT&T stock (Federal Communications Commission 1939, 88). By 1907 the Baker-Morgan group had gained effective control of the management of the company and returned Vail as president. The return of Vail, supported by the Baker-Morgan interests, marked a change in strategy that would have momentous consequences for competition in the industry, and in particular, for the independents. AT&T slowed its investment in internal expansion (the licensing of local Bell affiliates) in order to launch an ambitious campaign of external acquisitions (Gabel 1969, 351). With the financial support of the banking group, Vail began to buy up the independent competitors to the Bell System.

Vail's return inaugurated another important shift in policy: Western Electric began to sell equipment to the independents. This initiative was the strategic counterpart to the acquisition strategy. AT&T hoped to reduce the costs of eventually integrating the independents by ensuring that they were already using Bell System equipment.

The acquisition strategy raised the specter of government intervention under antitrust law. In 1888 the Interstate Commerce Commission (ICC) had been given the authority to regulate telegraph services provided by companies that had benefited from federal subsidies (Herring and Gross 1936, 210). In 1910 the Mann-Elkins Act amended the Interstate Commerce Act (1887) to grant the ICC the power to regulate the telephone companies. However, there was confusion concerning which companies were subject to federal regulation. In 1913 the independents complained to the attorney general that AT&T was acting unfairly to eliminate competition. The attorney general referred the matter to the ICC, which ordered an investigation (65). In order to deflect this attack, AT&T agreed to a settlement that later became known as the Kingsbury Commitment, named for the attorney who negotiated the agreement on behalf of AT&T (U.S. Letter from Nathan C. Kingsbury 1913). Under the terms of the Kingsbury Commitment of 1913, AT&T agreed to the following: (1) to divest itself of its 30 percent interest in the telegraph company Western Union, (2) to stop acquiring competing independent telephone companies, and (3) to connect the remaining companies to its long distance network (Kellog, Thorne, and Huber 1992, 16, 200). Also, AT&T would strive to balance its acquisition of non-competing companies with the sale of its own properties to the independents. This understanding proved to be a smokescreen for further acquisitions. Gabel notes that between 1913 and 1917, "the Bell System purchased over 241,000 stations from the independents and sold 58,000 stations" (1969, 353). Although AT&T began to liberalize its connection policy, the toll revenues derived from interconnection were not divided equitably with the independents. Specifically, the formula for sharing toll revenues did not recognize the vital contribution that local facilities made to the provision of long distance. As a result, the independents did not receive adequate compensation for the local facilities that initiated and terminated long distance calls in their territories. This made them less profitable than the Bell System companies and, ultimately, less appealing to investors.

As AT&T regained its dominant position in the industry, its activities were increasingly subject to public scrutiny. The Kingsbury Commitment was an early attempt to reach a negotiated settlement with regulators on terms that were favorable to the expansion of AT&T's core business: public telephony. But this was largely a defensive reaction to public concerns. Perhaps the most important legacy of Vail's second mandate was his endorsement of economic regulation. Gerald Brock notes that Vail had begun to embrace the possibility of economic regulation as early as 1907. Brock quotes the 1907 Bell System annual report as stating: "It is contended that if there is to be no competition, there should be public control. It is not believed that there is any serious objection to such control, provided it is independent, intelligent, considerate, thorough and just, recognizing, as does the Interstate Commerce Commission in its report recently issued, that capital is entitled to its fair return, and good management or enterprise to its reward" (1981, 158). Moreover, Brock suggests that Vail's support for regulation may have been reinforced by the experience of railroad regulation. At that time, the ICC was responsible for regulating the U.S. railroad industry. From this experience Vail would have concluded that regulation was compatible with profits.

Vail began to articulate a new corporate philosophy that he hoped would soon become public policy. It was based on the rallying cry "One system, one policy, universal service." Alan Stone has noted that Vail was simply capitalizing on the dominant public philosophy of the period, progressivism, which favored public control of large business (1997, 31). The new corporate philosophy was open to government regulation under certain conditions. AT&T would strive to achieve universal service and would submit to price regulation, if the government would accept the principle that telephone service should be offered non-competitively. From Vail's point of view competition and regulation were incompatible. If AT&T accepted the regulation of its prices, then it expected the government to protect it from competition. By submitting to regulation AT&T would gain support for its monopoly. From the government's point of view, a stable market, committed to the principle of universal service, was a welcome alternative to uneven and disorderly competition. For most of this century Vail's slogan "One system, one policy, universal service" reflected the prevailing consensus among consumers, regulators, and telephone providers on how the industry should be organized and operated.

What were the implications of Vail's strategy for competition with the independents? It is generally acknowledged that the year 1907 marked the highpoint of competition in the telephone industry (Brock 1981, 174). Almost half the telephones in operation in the United States in that year were owned by independent telephone companies. However, as a result of Vail's strategy, AT&T effectively regained its dominant position by the mid-1920s.

Attempts by the independents to establish a national alternative to the Bell System were thwarted by a number of factors. First and foremost, Vail's strategy, particularly following his return in 1907, proved to be very effective. But a large measure of this success was due to a failure of public authority. The Justice Department and the ICC were singularly ineffective at enforcing the Kingsbury Commitment. Nor were they successful in compelling AT&T to implement a more favorable division of long distance revenues. However, the independents were also to blame as they failed to organize themselves to compete with Bell at a national level. In 1897 the independents formed the National Association of Independent Telephone Exchanges (Herring and Gross 1936, 63). One of the objectives of the association was the creation of a long distance network. The independents, however, were more successful in complaining to government about AT&T's anti-competitive practices than they were in mobilizing to provide an alternative network of long lines. It warrants mention that the independents were not simply competitors to AT&T. In many cases they were also competitors to each other. But there is also evidence to suggest that their attempt to establish a long lines network was hampered by external factors. It appears that an informal agreement existed among the large banks not to finance them during a period when these banks were competing with each other for control of the gas and electric utilities in

New York City (Federal Communications Commission 1939, 88). Finally, it should be noted that not all the independents were opposed to the idea of being acquired by AT&T. For many owners, a buyout would be very profitable.

In 1921, with the support of AT&T and the independents, Congress passed the Willis-Graham Act. The act authorized the merger or acquisition of competing companies provided the consolidation had been reviewed and approved in advance by the ICC. This effectively abrogated the Kingsbury Commitment. Following the passage of the Willis-Graham Act, AT&T accelerated its acquisition of independent companies. But for all intents and purposes it had already become impossible for the independents to compete with the Bell System by the time of the act's passage in 1921 (Federal Communications Commission 1939, 141).

It was previously noted that AT&T withdrew from the domestic telegraph industry in 1914 when it sold its interest in Western Union as part of the Kingsbury Commitment. This was not AT&T's only incursion into an adjacent market. As early as 1907, Vail took an interest in the three-element vacuum tube invented by Lee De Forest (1873–1961). Vail was concerned that radio-communication could be used to establish a competitor to Bell's wire-based long lines (Federal Communications Commission 1939, 189). AT&T responded to this potential threat by gaining control of the patents on the three-element vacuum tube and by launching an ambitious program of research designed to fully explore the potential of this technology (190).

There was energetic competition for control of radio patents between the major manufacturers of electric equipment: General Electric, Westinghouse, and Western Electric. This had the effect of retarding innovation during World War I, as no one was able to exploit these technologies without potentially infringing upon another company's patent (Stone 1997, 38). In 1919 the government encouraged the creation of the Radio Corporation of America (RCA), a corporation with close ties to General Electric. In 1920 the government stepped in to create a patent pool for the sharing of radio patent rights between the major manufacturers.

The creation of the patent pool facilitated the creation of the first broadcasting stations in the early 1920s. AT&T attempted to enter the broadcasting market using a common carrier model that it referred to as "toll broadcasting." AT&T would own and operate the facilities, but it would not provide content. Programming companies could contract with AT&T for the use of its channels. AT&T's broadcast competitors opposed the company's foray into broadcasting and, eventually, were successful in mobilizing public opinion against the company. In 1926 it sold its broadcasting assets to RCA.

Stone (1997) has noted that AT&T's corporate structure was essentially in place as early as 1899 (25). Although this was the case, it was not until the early 1920s that the company had conclusively reversed the fortunes of the independents through acquisitions. During the early decades of the century, AT&T

also moved internally, within the Bell System, to consolidate its control of the telephone industry. It did this by increasing its share in the ownership of the operating companies that were part of the Bell group. In tandem with its success in promoting the philosophy of "one policy," whose goal would be "universal service," AT&T, with the assistance of its bankers, was strengthening the ties of ownership that would bind together its "one system." By 1934 the intercorporate relations between the principal Bell System companies were the following:

1. AT&T (holding company with controlling securities in the operating companies and the operating company for long lines operations)
2. The twenty-one operating companies throughout the United States (AT&T had 100 percent of the voting stock in ten companies, more than 99 percent of the voting stock in six companies, more than 65 percent in three companies, and less than 50 percent in two companies)
3. Western Electric Company (99.18 percent owned by AT&T)
4. Thirty long lines companies (100 percent owned by AT&T)
5. Bell Telephone Laboratories (50 percent owned by AT&T, 50 percent owned by Western Electric)
6. Bell Telephone Company of Canada (24.22 percent owned by AT&T) (Federal Communications Commission 1939, 66)

1.4 THE EARLY REGULATION OF U.S. TELECOMMUNICATIONS

Federal regulation of communications began with the enactment of the Postal Roads Act on July 24, 1866. The act provided rights-of-way to telegraph companies for the construction of lines over public domain lands and over postal roads. These rights were granted on the condition that companies register with the postmaster general (Herring and Gross 1936, 210). As noted previously, in 1888 the ICC was given authority to regulate telegraph services provided by railroad and telegraph companies that had benefited from subsidies in land or credit. Essentially, the ICC was empowered to ensure both the interconnection of telegraph lines and the fair exchange of messages between companies, and to prevent discrimination in the handling of messages.

In 1910 the Mann-Elkins Act extended ICC's authority to include the regulation of the rates and practices of companies involved in interstate communications. The act obliged common carriers "to provide service at just and reasonable rates, without unjust discrimination or undue preference" (Kellog, Thorne, and Huber 1992, 80). Moreover, the ICC had the power to invalidate unreasonable rates. In actuality, the ICC proved to be a relatively passive regulatory instrument. The Mann-Elkins Act did not require the mandatory posting of telephone and telegraph rates. Between 1910 and 1934 only four telephone and eight telegraph rate complaints were brought before the

ICC. In 1921 the Willis-Graham Act extended the jurisdiction of the ICC to include responsibility for overseeing acquisitions and consolidations.

In 1927 Congress enacted the Radio Act, which established the Federal Radio Commission. The commission was given responsibility for licensing radio stations and assigning frequency spectrum. More important from a telecommunications perspective, the act prohibited the cross-ownership of telephone and broadcasting companies (Kellog, Thorne, and Huber 1992, 19–20).

The Communications Act of 1934 created the Federal Communications Commission (FCC) with authority to regulate both broadcasting and telecommunications. The FCC inherited many of the powers and regulatory principles that had governed the ICC and the Federal Radio Commission.

It has been noted that the ICC was not an active or very effective regulator of the telephone industry. In part, this was due to the fact that most telephone traffic during this period occurred at the level of the local exchange; therefore, it was the responsibility of state regulators. Stone notes that by 1920 forty-five of the forty-eight states had public utility commissions with the power to regulate intrastate telephony (1997, 32). Subscribers dissatisfied with telephone rates had to take up these complaints with regulatory bodies at the state level.

One of the defining characteristics of the U.S. regulatory landscape has been the cohabitation of state and federal regulation of telephone companies. Also known as two-tiered regulation, this has meant that in theory, and very often in practice, a company will be subject to two levels of government control. This would produce jurisdictional clashes between federal and state levels of authority because in many cases it is virtually impossible to separate telephone operations on the basis of which facilities (or services) are used for interstate communication and which are used for intrastate communication. For example, the local exchange facility is the gateway to both local and long distance telephony. Generally, the courts interpreted the U.S. Constitution in a way that favored federal preemption (overriding) of state authority where a case could be made that state regulation prevented a federal agency from executing its mandate. As we will see in the next chapter, the Canadian approach divided jurisdiction on the basis of whether a *company* fell under either federal or provincial authority, thus precluding two-tiered regulation. The U.S. model relied, and still relies, on a sectorial approach, which divides economic activity into three domains: intrastate, interstate, and international. The FCC's authority encompassed the regulation of *interstate* and *international* telecommunications. Intrastate communications was the domain of state regulation as the Communications Act of 1934 inherited this framework from the Interstate Commerce Act of 1887.

Under the Communications Act of 1934, the obligations of common carriers were stipulated in section 201. (a), which stipulated that a carrier must provide service "upon reasonable request" and that "all charges, practices, classifications, and regulations for and in connection with such communication service, shall be just and reasonable." There was also a prohibition against discriminatory behavior. Section 202 stated the following: "It shall be unlawful for any common carrier to make any unjust or unreasonable discrimination in charges, practices, classifications, regulations, facilities, or services for or in connection with like communication service directly or indirectly, by any means or device, or to make or give any undue or unreasonable preference or advantage to any particular person, class of persons, or locality . . ." Every common carrier was required to file a public schedule of tariffs with the FCC showing all charges and "all classifications, practices, and regulations affecting such charges" (section 203. (a)). Notice of changes to rates or new rates had to be filed ninety days before they went into effect. The U.S. approach, unlike its Canadian counterpart, did not require FCC approval of all new or revised charges. Rather, the FCC at its own discretion, or upon receiving a complaint, had the power to conduct a full hearing into the proposed charges. Following such a hearing, the FCC could then authorize or vary the proposed charges. It had twelve months to conduct a hearing, and if the hearing was not conducted within that time period, the rates would go into effect. In the event that the FCC sided against the company, it could order a refund to consumers (section 204. (a) (1)).

The Communications Act of 1934 granted the FCC almost unlimited powers to suppress or allow competition (sections 201 and 214; Kellog, Thorne, and Huber 1992, 19). In effect, under the 1934 act it was up to the FCC to determine whether monopoly or competitive conditions would prevail. It merits noting that for most of the century the FCC used its powers to conserve the monopoly of the Bell companies. As we shall see in chapter five, in the 1960s the FCC began to reconsider its approach in order to gradually introduce competition.

The Communications Act of 1934 charged the FCC with a number of oversight powers that impinged, in varying degrees, on the prerogatives of management. Carriers were required to file with the FCC all contracts they entered into with other carriers or with non-carriers. The act empowered the FCC to examine these transactions to determine if they had an impact on rates or services (section 215). Section 212 made it illegal for common carriers to form interlocking directorates without FCC approval. The commission also had the power to regulate mergers and consolidations (section 221. (a)).

An appeal of FCC decisions could be made in one of the circuit courts of appeal. The procedure for judicial review would eventually be covered by the Administrative Procedures Act (1946). When reviewing a decision of a regulatory tribunal, the court is "supposed to decide whether the agency has acted within its delegated authority and on the basis of a reasoned consideration of the evidence before it" (Bruce, Cunard, and Director 1986, 166). In theory, the courts were supposed to defer to the regulatory agency in recognition of its implicit expertise. However, as we will see in chapters five and six, on numerous occasions the courts substituted their own judgment for that of the regulatory agency only to be admonished by the U.S. Supreme Court (166).

The FCC had the power to enforce its decisions either through its own procedures or through the courts. Parties who claimed injury based on a carrier's failure to implement an order by the FCC were required to first file a complaint with the FCC. The U.S. act, unlike its Canadian counterpart, which we will consider in the next chapter, specified the amount of the fines that could be levied in cases of non-compliance. For example, willful violations of the act could be punished by fines of not more than $10,000 and imprisonment up to one year (section 501).

1.5 CONCLUSION

By the mid-1930s the telecommunications industry in the United States was highly concentrated with AT&T as the dominant player in telephony and Western Union as the dominant player in the telegraph industry. Although Western Union was subject to a measure of direct competition, this was hardly the case with AT&T.

The telegraph industry was an oligopoly consisting of two groups, Western Union and the International Telephone and Telegraph (ITT) system. Together these groups accounted for 99.89 percent of total operating revenues for domestic and international telegraphy (Herring and Gross 1936, 183). Both groups provided service throughout the United States and overseas. Western Union was by far the larger company with operating revenues that accounted for 75.26 percent of total industry revenues. ITT accounted for 24.63 percent (184). Although the industry was competitive, competition was not vigorous. Rather it was typical of the type of limited competition found in oligopolistic markets.

The industry structure for telephony was even less competitive. The long distance voice market was a monopoly controlled by AT&T. Nor was there any direct competition in the local market: the local telephone market had been segregated into territories that were controlled by either the Bell System or the independents with no competition within the boundaries of these territories. In their 1936 study James Herring and Gerald Gross note that in 1932 "the companies comprising the Bell System owned 93.2 per cent of the total miles of wire and 84.7 per cent of the total number of telephones for these systems. They originated 86.1 per cent of the total number of calls originated; they received 90.5 per cent of the total exchange revenues and 93.6 per cent of total toll revenues . . ." (181). Moreover, AT&T's dominance of the industry was not limited to the business of common carriage; its manufacturing affiliate, Western Electric, was by far the most important manufacturer of telephone equipment. By 1934 Western Electric's sales accounted for 91.6 percent of this market (Federal Communications Commission 1939, 145).

2

Telegraphs and Telephones: Building the Canadian Telecommunications Mosaic, 1846–1946

2.1 INTRODUCTION

Commercial telegraph and commercial telephone service were introduced concurrently to the United States and Canada. But the industrial organization of the telecommunications sector in the two countries would differ. The term "industrial organization" refers to the economic and institutional structure of a sector. For the purpose of this book, it encompasses such factors as: the number of firms operating in the sector (which is an indicator of economic concentration, also known as horizontal integration), the ownership (whether public, private, or mixed), the extent of vertical integration (do companies provide inputs to their core business; e.g., in the case of telecommunications, whether common carriers manufacture and provide telephone sets to subscribers or manufacture their own switching equipment), the existence of barriers to entry, the degree of product differentiation, and the mode of regulation (the extent and type of government controls on outputs [quality control] and prices). As we shall see in this chapter, although the technologies were the same, their exploitation in Canada would follow a path distinct from that of the United States, producing a different industry structure. This would be particularly the case in the following areas: the application of patent law, the manner in which regulatory authority would be divided between provincial and federal levels of government, and the structures of ownership.

2.2 THE ORIGINS AND CONSOLIDATION OF THE CANADIAN TELEGRAPH INDUSTRY

In the United States the first commercial telegraph services were introduced by entrepreneurs who had acquired the rights to exploit Samuel Morse's patent. In

Canada the impetus to develop a commercial telegraph service came from a different source. Under the law of the time, patents in British North America could only be held by British subjects. This meant that the Morse patent was not recognized in Canadian law. As a result there was no opportunity for independent Canadian entrepreneurs to benefit from exclusive use of the patent. Instead, the telegraph was promoted by businessmen who saw an opportunity to extend and improve their existing businesses.

The first commercial large-scale telegraph company in Canada, the Montreal Telegraph Company, received its charter from the federal government in 1847. It was a joint venture, financed by Montreal businessmen and sponsored by the Montreal Board of Trade (Babe 1990, 38). By the end of 1847, the company had installed a telegraph line extending from Toronto to Quebec City via Montreal (Department of Communications 1983a, 1). During the 1850s, the Montreal Telegraph Company expanded operations from five hundred to nineteen hundred miles of telegraph lines. By 1870 the company was operating 12,400 miles of telegraph lines. Robert E. Babe, the Canadian telecommunications scholar, has noted: "By 1865 most towns and villages, harbours on the lakes, and the lumbering districts of the Ottawa River and Eastern Townships of Québec were served" (1990, 45).

While the Montreal Telegraph Company was expanding in Canada, the telegraph industry continued to grow in the United States, attaining the status of a major North American industry. Local and regional telegraph companies found that they could increase the value of their networks to clients by connecting them to the networks of other companies in other regions. Moreover, by agreeing not to compete with one another in the same territory, and by excluding other companies from the connection agreements, companies were able to reduce their vulnerability to competition. These agreements between regional telegraph companies resulted in the formation of telegraph cartels (fig. 2.1). In 1857 the Montreal Telegraph Company joined the six-member North American Telegraph Association. The other principal members of the consortium across the border were the Western Union Company and the American Telegraph Company. The members of the cartel agreed that Montreal Telegraph would operate in the territories of the Canadas (Canada East and Canada West—i.e., Quebec and Ontario), as well as certain regions in some of the northern United States (Babe 1990, 46).

The Montreal Telegraph Company was so profitable that the business drew the attention of another group of investors who formed the Dominion Telegraph Company in 1968. Dominion Telegraph was soon operating stations at important points between Detroit and Quebec (Department of Communications 1983a, 1). But Dominion was shut out of the North American Telegraph Cartel. In order to offer service to points across the border the company negotiated an agreement to interconnect its network with that of the Atlantic and Pacific Telegraph Company of the United States.

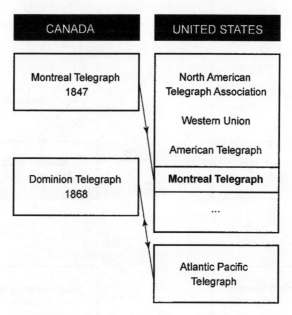

Figure 2.1   Development of a North American Telegraph Cartel

When Jay Gould gained control of Atlantic and Pacific Telegraphs in 1873 he set the company on a course of fierce competition with Western Union. In 1881, as a result of this competition, Western Union was forced to merge with Atlantic and Pacific Telegraphs and Gould had become a major stockholder in Western Union.

By 1891 Gould had achieved total control of Western Union in the United States. This would spell the demise of Dominion Telegraph as the Canadian company soon became a pawn in a grander strategy, orchestrated by Gould, to gain control of the Canadian telegraph market (fig. 2.2). Gould's designs on the Canadian market would also mark the beginning of difficulties for the Montreal Telegraph Company. Through its merger with Atlantic and Pacific, Western Union was able to establish commercial ties with Dominion Telegraph. It will be remembered that Dominion Telegraph depended on Atlantic and Pacific for links with the United States. Gould's first step was to weaken Montreal Telegraph. He did this by encouraging Dominion Telegraph to engage in a price war with Montreal Telegraph. The second step involved the purchase by Western Union of the Great North Western Telegraph (GNWT) Company, a collection of smaller companies that had been organized in Winnipeg in 1880.

In 1881, under pressure of a fierce price war with Dominion, Montreal Telegraph shareholders accepted an offer to grant GNWT a ninety-seven-year lease on their lines (Babe 1990, 50). In effect, they were transferring the operation of their network to GNWT. When Quebec and Ontario were added

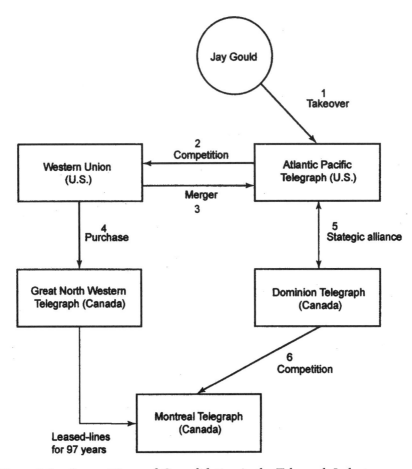

Figure 2.2   Competition and Consolidation in the Telegraph Industry

to Gould's existing ownership of several telegraph companies in Canada's Atlantic Provinces, Gould's near monopoly control of the Canadian market extended from Ontario to the Atlantic coast.

   We will now turn to the development of the telegraph in western Canada. In 1869 the Dominion of Canada purchased Rupert's Land from the Hudson Bay Company. (The territory had been granted to the company in 1670 by King Charles II of England.) Rupert's Land encompassed most of today's Prairie Provinces (Manitoba, Saskatchewan, and Alberta), northern Ontario, and northwestern Quebec. In 1870 the Canadian government acquired the Northwest Territories from Great Britain. At the time Ottawa was concerned that the United States would annex the western territories, including British Columbia. Therefore, the Canadian government, in order to strengthen its ties with the western territories and to discourage American incursions into them, decided to venture into the telegraph business. From 1872 to 1879 the federal government constructed thirteen hundred miles of lines from Thunder Bay in western Ontario to the city of Edmonton, east of the Rocky Mountains. In 1881 these lines were combined with other government owned facilities and became known as the Government Telegraph and Signal Service (Babe 1990, 55).

In the 1880s Gould's GNWT also began to turn its attention westward. A telegraph line was built linking the company's operations in Ontario to Winnipeg, but the company failed in its attempt to purchase government lines linking Winnipeg to Edmonton. It was during this period that opposition to Western Union's near monopoly in Canada began to develop in the Canadian Parliament. Although there were calls for a government takeover of the Western Union subsidiary in Canada, no action was taken by Parliament (Babe 1990, 56).

When British Columbia entered the Canadian federation in 1871 as Canada's sixth province, the federal government promised to complete a rail link with the province within ten years. The project, which was delayed by a political scandal (it was revealed that the Conservative Party had accepted $300,000 in campaign contributions from the firm that was granted the contract to build the railroad) and by an economic depression, began in earnest in 1880 when the government signed a contract with the Canadian Pacific Railway Company (Bercuson and Granatstein 1988, 37). Canadian Pacific was the beneficiary of massive government subsidies in cash and land. Moreover, when the federal government privatized its telegraph operations in western Canada, these were turned over to Canadian Pacific. To complement its western expansion, Canadian Pacific purchased rail lines extending east through Ontario and Quebec to the Atlantic Provinces. Eventually, it acquired the right to string telegraph lines along its railway installations. In November 1885 the transcontinental railway stretching from Halifax to Vancouver was completed with the driving of the last spike at Eagle Pass in British Columbia.

The completion of the transcontinental railway link coincided with the first coast-to-coast telegraph service based exclusively on the use of Canadian facilities. Earlier transcontinental Canadian service had depended upon connections with U.S. telegraph companies. Messages would be routed for part of their journey through the United States and then back to Canada. In 1886 Canadian Pacific Railway Telegraphs was officially organized as a subsidiary of the Canadian Pacific Railway, even though it had already been in operation for five years.

Two other railways with national aspirations would come to play a roll in the growth of the telegraph industry: Canadian Northern Railway and Grand Trunk Pacific Railways. Canadian Northern began as a small Manitoba railway. Founded in 1899 it was popular with grain farmers in the Prairie Provinces. In the years preceding the First World War, Canadian Northern extended its service eastward to Quebec City and westward to Vancouver. After incurring heavy construction debts, Canadian Northern Railway was national-ized in 1917. Grand Trunk Pacific Railway was intended to be a national com-petitor to Canadian Pacific Railway. Operating in eastern Canada, it was a subsidiary of the Grand Trunk Railway of Canada, largely owned by British in-vestors. Its lines were to run west from Winnipeg to British Columbia; from Winnipeg its lines were to be connected to eastern Canada through the gov-ernment owned line, the National Transcontinental Railway.

By 1915 the principal telegraph companies operating in Canada were: Canadian Northern Telegraphs, Grand Trunk Pacific Telegraphs, and Cana-dian Pacific Telegraphs. All were subsidiaries of their parent railway compa-nies. Gould's plan to control the Canadian telegraph industry had been thwarted by the emerging partnership between the telegraph and railroad business. GNWT, which Gould had set up to compete with Canadian Pacific Telegraphs for the western telegraph market, found that it was unable to com-pete with the telegraph subsidiary of the Canadian railway company. In 1915 its operations were acquired by Canadian Northern. The demise of GNWT was to a large extent due to the Canadian subsidiary's inability to integrate an emerging news wire business with its telegraph operations.

There was a natural synergy between the telegraph and wire service busi-ness at the turn of the century. In many locations the telegraph operator acted as a pseudojournalist: Not only would he relate the news he received over the wires to his local community but he would also gather local news for distri-bution to other points. It was not long before a new business emerged, the news wire. These news wire services became a major source of revenue for the telegraph industry. Canadian Pacific Telegraphs was able to outperform Gould's GNWT in large part because it was able to procure the exclusive Canadian rights to the Associated Press wire service, which emanated from the United States. When GNWT was shut out of this business, it was de-prived of an important source of revenue. It would not be long before these exclusive arrangements between telegraph carriers and wire service (content) providers would be challenged as a form of unfair competition.

Babe notes: "For years thereafter virtually every daily in Canada as a prac-tical matter had to subscribe to both Canadian Pacific's news service and its transmission service, simply because the company bundled them together under one flat price" (1990, 58). When a competitive service, the Western Associated Press, emerged in Winnipeg in 1907, Canadian Pacific terminated special press rates for telegraph to the city. Complaints about these and other abuses led to a Board of Railway Commissioners ruling that Canadian Pacific Telegraph's press rates had been unduly discriminatory and were, therefore, unlawful. Following this ruling Canadian Pacific withdrew from the wire serv-ice business. This action brought an end to the vertical integration of content (information production) and carriage (the public transmission of informa-tion) in the early decades of the telecommunications industry.

In the period following the First World War Canada's telegraph industry was consolidated. This trend was a logical consequence of the consolidation of the railroad industry during the same period. In 1919 the federal government established Canadian National Railways as a publicly owned, crown corporation. Between 1919 and 1923 Canadian National Railways acquired a number of railways (fig. 2.3). These included Canadian Northern Railways, Grand Trunk Railways, and Grand Trunk Pacific Railways. The telegraph operations of Canadian Northern (which already operated a national service in competition with Canadian Pacific) and Grand Trunk Pacific were consolidated to form the Canadian National Telegraph Corporation. On January 1, 1921, Canadian National Telegraphs came into being. Thus by the mid-1920s the telegraph industry had been reduced to a duopoly, controlled by Canadian interests, with strong corporate ties to the Canadian railway industry.

The regulation of telegraphic communication began with the Telegraphs Act of 1852. The act empowered telegraph companies incorporated under the Federal Companies Act (now the Canada Corporations Act) to construct lines along public roads and highways, or across or under navigable waters. It required that messages be transmitted in the order received (in order to prevent discrimination). The act authorized the government to temporarily requisition the use of

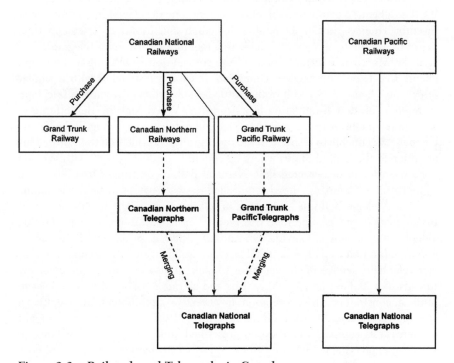

Figure 2.3   Railroads and Telegraphs in Canada

telegraph facilities, or if so required, expropriate them with compensation. In 1881 the act was amended to include an obligation to maintain the secrecy of messages. These provisions of the Telegraphs Act would also apply to the public telegraph operations of the railway companies that had been incorporated under special acts of Parliament.

2.3 THE ORIGINS AND CONSOLIDATION OF THE CANADIAN TELEPHONE INDUSTRY

As noted in chapter one, Alexander Graham Bell received his patents for the telephone in 1876 and 1877. Bell's father, Melville Bell, was a Canadian resident. As a gift to his father, Alexander Bell transferred three-quarters of the rights to exploit the patent in Canada to the elder Bell. The remaining one-quarter interest in the patent was transferred to a Boston businessman, Charles Williams (Babe 1990, 66). In February 1879 Melville Bell granted Dominion Telegraph Company of Canada a five-year license to work the patent throughout the country. (The license excluded Toronto and Hamilton where other Canadian agents had already been authorized to exploit the patent.)

The Montreal Telegraph Company entered the telephone business by obtaining the rights to a competing telephone patent, one developed by the American inventor Thomas Edison. Dominion Telegraph and Montreal Telegraph then began to use the telephone as a marketing tool in their battle for telegraph subscribers. Montreal Telegraph initiated the competition by offering its telegraph clients free telephone service. Dominion Telegraph responded with a similar offer to its clients. As a result of the battle for subscribers, Dominion Telegraph suffered a substantial reduction in its earnings. When Melville Bell offered to sell the telephone patent outright to Dominion Telegraph in August 1879, the company was unable to respond to the opportunity. Melville Bell then turned his attention to the United States where he concluded a deal in 1880 with William Forbes, president of National Bell. National Bell, reconstituted in 1880 as the American Bell Company, already controlled the U.S. patent to Bell's invention.

While Melville Bell was attempting to sell the rights to the telephone patent to American interests, a petition was filed in Parliament to incorporate the Bell Telephone Company of Canada. On April 28, 1880, a special act of Parliament approved the Bell Canada Charter. The charter empowered the company to provide telephone service to *all* regions of Canada. It also authorized the company to construct its lines along public roads and all public rights-of-way, subject to minimal constraints. The corporation, however, was prohibited from operating a telegraph service, and it was required to return to Parliament each time it sought to increase its capitalization, beyond prescribed limits, through the issuance of stocks or bonds (Babe 1990, 68). Babe notes that the parliamentary record does not suggest that Parliament was aware of an impending American takeover of the emerging Canadian telephone industry, that is, of Melville Bell's efforts to sell the rights to the patent to the Americans. It is also unclear whether Parliament was ever informed that the initiative to incorporate Bell Canada was actually sponsored by agents of the American Bell Company, that is to say, whether or not Parliament was deliberately deceived (69–70).

Although the Bell Telephone Company of Canada was authorized to operate a telephone service throughout Canada, it would never fully exercise its prerogative. Instead, the company limited its operations to population centers where its service would be most profitable. The introduction of telephone service to other, less profitable areas was left to its competitors. The Bell Telephone Company invested heavily in the southern corridor of Quebec and Ontario, which encompassed Montreal, Ottawa, Kingston, Toronto, Hamilton, and Windsor. Bell's expansion into this corridor was achieved primarily through the acquisition of the existing telephone facilities of Dominion Telegraph and Montreal Telegraph. The company also established operations in the Atlantic Provinces and in western Canada. But these latter operations were done on a more limited basis.

In the mid-1880s the Bell Telephone Company began to limit its operations in the Atlantic Provinces. At the time, Bell's ownership of telephone installations in the region was not very extensive. It consisted of telephone exchanges with only several hundred subscribers. In 1885 Bell sold its telephone assets on Prince Edward Island. This led to the creation of the Island Telephone Company. In 1888 Bell sold its facilities in Nova Scotia and New Brunswick to the Nova Scotia Telephone Company. In his history of Bell Canada, Jean-Guy Rens notes that the sale took place for $55,000 in liquidity, $65,000 in stocks, and a commitment to purchase equipment manufactured by Bell over its competitors given comparable prices (1993a, 142). At the time of the sale the network consisted of 539 subscribers served by four telephone exchanges (142). In 1889 Bell bought back the exchanges in New Brunswick and resold them to the newly created New Brunswick Telephone Company. The network was sold for $50,000, half in cash and half in stock, which was equivalent to a 31 percent stake in the ownership of the company (143). On July 1, 1911, Maritime Telegraph and Telephone Company (incorporated in 1910) purchased the entire undertaking of the Nova Scotia Telephone Company (Department of Communications 1971a, 15). In 1962 Bell acquired shares of the Avalon Telephone Company of Newfoundland. On January 1, 1970, Avalon's name was changed to the Newfoundland Telephone Company (11). Newfoundland Telephone was the principal telephone company operating in the province of Newfoundland.

Bell's strategy of indirect control through the partial ownership of provincially based telephone companies was successful. For example, in 1992 Bell Canada Enterprises had a 31.4 percent interest in New Brunswick Telephone, a 33.8 percent interest in Maritime Telephone and Telegraph, and through Maritime Telephone, an interest in Maritime Telephone's Prince Edward Island subsidiary, Island Telephone. Bell Canada Enterprises has maintained a 55.7 percent interest in NewTel, the parent of Newfoundland Telephone (Communications Canada 1992, 17).

When Bell reduced its ownership of telephone operations in the Atlantic Provinces in the 1880s, its interests were not very extensive. Moreover, the transfer of ownership took place between privately owned corporations. When Bell withdrew from its operations in the Prairie Provinces of Manitoba, Saskatchewan, and Alberta in the early 1900s, it did so under a different set of circumstances: (1) Bell's operations were more extensive and entrenched in this region than they had been in the Atlantic Provinces, and (2) the change in ownership involved a transfer from private enterprise to publicly owned corporations controlled by the provincial governments.

It is important to note that in 1885 Canada's patent commissioner ruled that Bell's original patent was void. Contrary to Canadian patent law, the Bell Telephone Company was not using the patent to manufacture equipment in Canada. Instead, it was importing it from the United States. The voiding of the Bell patent opened the way to competition in local telephone service. Municipalities in the Prairie Provinces began to set up telephone exchanges in areas that were not adequately served by the Bell network.

The Bell Telephone Company began operations in Winnipeg, Manitoba, in 1881 when it took control of a number of small, local, private companies (Department of Communications 1971a, 40). Bell then extended its service to the major centers of Winnipeg, Portage la Prairie, and Brandon. In 1889 the Manitoba legislature passed a law empowering municipalities to own and operate local telephone exchanges. The loss of Bell's basic patent meant that the company was subject to competition in these markets. In order to quash competition Bell turned to a new strategy: It refused to connect the local telephone exchanges, which were operated by the municipalities, to its long distance network. As a result, the growth of telephone service in the province was impeded. In 1906 and 1908 the province of Manitoba responded with legislation authorizing the government to construct its own long distance facilities. Finally, on January 15, 1908, the province purchased Bell's telephone operations outright for $3.3 million (Babe 1990, 103). As a result, the provincial government of Manitoba became the principal operator of the telephone network. Rens notes that the Manitoba system became "the first entirely state owned telephone network in North America" (1993a, 188, our translation).

On May 1, 1909, the government of Saskatchewan acquired Bell's operations, which served sixteen hundred subscribers. This was followed two months later by the government's purchase of the Saskatchewan Telephone Company. The Saskatchewan approach did not involve total control of the telephone network as a provincial monopoly. The provincial government was content to be the sole operator of the long distance network; it limited its control of local telephone companies to those in the larger cities. Although it offered to support local municipalities attempting to establish exchanges, the provincial government did not adopt a policy of total ownership (Babe 1990, 107).

In 1887 the mayor of Calgary, Alberta, solicited Bell Telephone of Canada to set up a telephone exchange in the city. After accepting the invitation, Bell quickly established local telephone service in Calgary, Lethbridge, and Medicine Hat, and a toll service that included lines from Calgary to Edmonton (Department of Communications 1971b, 53). In 1905 Alberta entered the Canadian federation. In January 1905, before Alberta joined the confederation, the city of Edmonton purchased the lines and plant of the Edmonton District Telephone Company. In 1906 one of the first acts of the new Alberta legislature involved the telephone industry. In a pattern already well established in the other western provinces, the legislature empowered municipalities to own and operate local telephone exchanges. In the same year Alberta Government Telephones was formed as a part of the provincial Department of Public Works. In July 1907 the government began to construct a long distance network to areas not served by Bell. By the end of the year the provincial government had installed twelve telephone exchanges and purchased a number of independent telephone companies. Finally, on April 1, 1908, the government of Alberta purchased all of Bell's assets in the province. Alberta Government Telephones would provide telephone service throughout the province. But its territory did not include Edmonton, where the local municipality maintained its ownership of telephone operations. In 1958 the Alberta Government Telephones Act transformed the provincial telephone company into an independent crown corporation, which enabled the corporation to obtain financing on the open market.

The Vernon and Nelson Telephone Company was incorporated in British Columbia on April 30, 1891. The act of incorporation authorized the company to operate within British Columbia and to establish connections with points outside the province (Department of Communications 1971b, 61). In 1904 the company purchased the assets of the Victoria and Esquimalt Telephone Company (incorporated in 1880) and the New Westminster and Port Moody Telephone Company (incorporated in 1884). On July 5, 1904, the company changed its name to the British Columbia Telephone Company (B.C.Tel), Limited. Unlike the other western provinces, British Columbia did not enact legislation authorizing municipalities to operate local telephone service (Babe 1990, 112). However, independent telephone companies did establish service in areas not served by the dominant carrier. Although there was pressure to provincialize the British Columbia Telephone, these plans were never realized.

Having rejected the option of public ownership, the provincial government began to consider other means to regulate the industry. When legislation to regulate the company was presented to the provincial legislature in 1912, British Columbia Telephone responded by petitioning the federal government to incorporate a new company, the Western Canada Telephone Company. In doing so the company hoped to escape provincial regulation of its business. But the move did not achieve its purpose. Although the federal government granted the petition, it unexpectedly placed the new corporation under the regulatory authority of the federal Board of Railway Commissioners. British Columbia Telephone was able to evade federal regulation for several years by delaying the transfer of its assets to the new company. Matters were confused further when Western Canada Telephone Company changed its name to British Columbia Telephone Company (without using the term "limited"). In 1923 the two companies were consolidated, thus resolving the regulatory question. Thereafter, British Columbia Telephone was to be regulated by federal authority (Babe 1990, 112).

In the 1920s British Columbia Telephone was sold to the National Telephone and Telegraph Corporation of the United States, later known as the Anglo-Canadian Telephone Company of Montreal. In 1955 Anglo-Canadian became a subsidiary of the General Telephone and Electronics Corporation (GTE), whose headquarters were (and still are) located in Stamford, Connecticut. By the early 1990s Anglo-Canadian was the majority shareholder in British Columbia Telephone with 50.2 percent of shares. It also held 50.7 percent of shares in Québec-Téléphone, the largest independent telephone company operating in Quebec with 254,500 subscribers (Communications Canada 1992, 17; Association des compagnies de téléphone du Québec 1991, 7). British Columbia is the only province with a dominant telephone company controlled by foreign interests.

Historically, Ontario and Quebec have been the commercial and industrial dynamos of the Canadian economy. It, therefore, made good business sense for Bell Canada to concentrate its efforts on expanding and protecting its market in these two provinces. Bell's competitors were the independent telephone companies that sprang up throughout Ontario and Quebec. Babe's research into the early history of the telephone industry in Ontario reveals that 760 independent telephone companies were started in Ontario alone between 1906 and 1920 (1990, 117).

Following its incorporation in 1888, Bell employed a variety of methods to maintain its dominant position against the incursions of the independent telephone companies. Initially, the company relied on its control of the Bell patent. However, as we have seen, this protection was short-lived. The Special Act of Parliament incorporating Bell Canada (1880) was a source of considerable competitive advantage, because it guaranteed the corporation's right to operate a long distance network. This privilege was not automatically accorded to Bell's competitors, the independent phone companies. Other weapons in the Bell arsenal included acquisitions, prohibitions on interconnection with its network, predatory pricing practices (i.e., pricing below cost to drive out competition), and recourse to regulatory tribunals. These practices closely paralleled those employed by AT&T in the United States.

In 1931 Canada's largest telephone common carriers joined together to form a cooperative, unincorporated consortium for the purpose of developing and maintaining an all-Canadian long distance network that would operate from coast-to-coast. The consortium would also work to establish uniform operating procedures in order to facilitate the handling of long distance calls and other traffic such as television and radio programs, data, and defense communications (Department of Communication 1971b, 4). The initial members of the consortium, known at the time as the Trans Canada Telephone System (TCTS), were:

The Bell Telephone Company of Canada (later Bell Canada)
The Maritime Telegraph and Telephone Company
The New Brunswick Telephone Company
Manitoba Government Telephones (later Manitoba Telephone System)
Saskatchewan Government Telephones (later Saskatchewan Telecommunications)
Alberta Government Telephones
British Columbia Telephone Company

In 1975 TCTS changed its name to Telecom Canada. In 1992, as the result of reorganization, it became known as Stentor Canadian Network Management.

The years preceding the Second World War were punctuated by a number of significant events. These included: (1) the first coast-to-coast commercial radio broadcast using Canadian National lines (1925); (2) the inauguration by TCTS of a transcontinental long distance telephone network (1932); (3) the securing by Canadian National Telegraphs and Canadian Pacific Telegraphs of the network contract to transmit programming prepared by the newly created Canadian Radio Broadcasting Commission—this marked the beginning of the Canadian Broadcasting Network (1932); and (4) the first nationwide weather gathering service, provided by Canadian National and Canadian Pacific (CNCP) Telecomunications (1939) (Department of Communications 1983a, 3).

2.4 THE EARLY REGULATION
OF CANADIAN TELECOMMUNICATIONS

The Board of Railway Commissioners for Canada was responsible for regulating the telephone and telegraph industries between 1906 and 1938. In 1938 this responsibility was transferred to the Board of Transport Commissioners (1938–1967), which was superseded by the Canadian Transport Commission in 1967 (1967–1976). In 1976 the Canadian Radio-Television Commission was rebaptized the Canadian Radio-television and Telecommunications Commission (CRTC) and given responsibility for regulating the companies that fell under federal jurisdiction.

Until the enactment of the Canadian Telecommunications Act of 1993, the obligations and privileges of the federally regulated, telegraph and telephone companies were spelled out in the Railway Act (1988).[1] It was the Railway Act that stipulated that telegraph and telephone tolls had to be just and rea-

sonable and that the companies could not discriminate with respect to customers or services (340. (1) (2)). Under the Railway Act, the CRTC was
granted considerable powers to ensure that these obligations were respected.
All tolls and all charges for using or leasing telephones or telegraphs had to
be approved by the commission.[2] Without exception, all tolls had to be filed
with the commission (335. (2)). Not only were tolls subject to commission
approval, but the commission was empowered (after consideration) to substitute its own tolls for those proposed by the companies. Having approved a tariff, the commission had the power to determine "as a question of fact"
whether the terms of the tariff were being respected; if necessary, it could
suspend or disallow a tariff found to be in violation (340. (4)). The commission was granted general powers to make orders with respect to all matters relating to traffic, tolls, and tariffs.

A second set of responsibilities dealt with the management prerogatives of the
regulated companies. The Railway Act stipulated that all contracts, agreements,
and arrangements that the companies entered into with other companies, with
provinces, or with municipalities were subject to commission approval (388.).
The companies were required to submit annual reports of their assets, liabilities,
capitalization, revenue, working expenditures, and traffic to the commission
(344. (1)). Moreover, the commission could require a company to furnish written statements showing its assets and liabilities, the amount of stock outstand-
ing, the consideration received for such stock, gross earnings and expenditures,
and other information of a financial nature (358. (1)).

The commission was granted a number of general powers to ensure that it
would be able to fully exercise its responsibilities. Until 1985 these were set
out in the National Transportation Act and the acts that preceded it. In 1985
the National Telecommunications Powers and Procedures Act (NTPPA) consolidated these powers with respect to telecommunications. For the most part,
the powers enumerated in the 1985 act were equivalent to those conferred
upon the regulatory bodies that preceded the CRTC. The NTPPA conferred
upon the commission a number of powers of *inquiry*. The act empowered the
commission to inquire into any complaint charging that a company had failed
to meet its obligations under the act (49. (1) (2)). Of its own initiative, or upon
request of the minister responsible, the commission had full powers to inquire
into, to hear, and to determine any matter under the act (52.). The commission was authorized to appoint any person to make an inquiry and to report
upon any application, complaint, or dispute. In addition to powers of inquiry,
the commission was granted powers of *enforcement* (84.). Under the NTTPA,
the commission was granted broad powers to make orders and regulations for
the purpose of carrying the act into effect (50. (1)). Most important, in the exercise of its jurisdiction the commission had all the powers, rights, and privileges as are vested in a superior court (49. (4)). This meant that its orders were

binding on parties covered by the act, and its findings upon any question of fact (such as the respect of tolls) were binding and conclusive (59. (3)).

A number of provisions were made for review or appeal of commission decisions. The commission had the authority to review and change its own orders and decisions (66.). The Governor in Council (i.e., the federal cabinet) also had the power to vary or rescind any order, decision, rule, or regulation of the commission (66.). Finally, orders or decisions of the commission could be appealed to the Federal Court of Canada. However, such appeals could only be made on questions of law or jurisdiction (68. (1)). There was no right of appeal to the federal court with respect to the substance of an order or decision—for example, on whether the commission had erred in its reasoning or with respect to the potentially harmful impacts of a decision.

Under Canadian law, the CRTC's jurisdiction (and the jurisdiction of regulatory bodies that preceded it) was defined in terms of the set of *companies* that fell under federal authority. For example, from the turn of the century to the late 1980s, the principal responsibility of the commission was the regulation of Bell Canada, B.C.Tel, and CNCP Telecommunications. The determination that a company fell under federal jurisdiction depended on the presence of two conditions. First, it had to be determined through an interpretation of the British North America Act (1867) that an enterprise fell under federal jurisdiction. Second, an enterprise had to be recognized as a "company" within the meaning of the Railway Act. In short, Canadian law did not acknowledge a priori a national, interprovincial sphere of economic activity as a jurisdiction subject to federal authority regulation. Rather, jurisdiction was granted to regulate specific companies with a national vocation as determined through a legal interpretation of the British North America Act and the Railway Act. Bell Canada, B.C.Tel, and CNCP were all federally incorporated companies. In large measure, it was the fact that these companies had been chartered as acts of Parliament that contributed to the determination (under the relevant acts) that they fell under federal jurisdiction.

2.5 CONCLUSION

By the late 1930s the basic structure of the modern, Canadian telecommunications industry had emerged. By this time most of the principal actors that would dominate the contemporary telecommunications industry had established their presence. Most important, a unique industry structure had developed.

The Canadian industry structure was unlike its counterpart in the United States, and unlike the Post Telegraph and Telephone (PTT) administrations that prevailed elsewhere in the world. This was particularly the case with respect to the ownership and regulation of the industry. This structure would define the

Canadian telecommunications industry in the period leading up to the liberalization and deregulation of the late 1980s and early 1990s. The principal features of the unique Canadian industry structure included:

1. A common carrier mosaic defined largely on the basis of provincial markets
2. A mixture of public and private ownership of the common carriers
3. A tradition of active government intervention in the form of acquisitions and subsidies, at both the federal and provincial levels, to ensure infrastructure development
4. A system of mixed and often conflicting regulatory jurisdictions, resulting from the mosaic, that would make it difficult to plan, coordinate, and create effective policy at the national level
5. Bell Canada's presence as the dominant carrier within the telecommunications mosaic
6. The influence of American markets and American ownership on Canadian industry structure

NOTES

1. The Railway Act (1888) underwent numerous revisions. We refer here to the act as it stood in 1988. Peter S. Grant has reproduced this version of the act in a collection of laws pertaining to the communications sector published in 1988 (Grant 1988).

2. In order to simplify exposition we will refer throughout this section to the commission (or the CRTC) as the agency responsible for regulating telecommunications. It should be noted, however, that the CRTC was created in 1968 to regulate broadcasting and was only assigned jurisdiction over telecommunications in 1976. From 1903 to 1938 the telephone industry was regulated by the Board of Railway Commissioners for Canada. From 1938 to 1967 the industry was regulated by the Board of Transport Commissioners. Finally, from 1967 to 1976 the Railway Transport Committee of the Canadian Transport Commission was responsible for regulating telecommunications under the Railway Act.

3

The Regulation of the Telephone Industry as a Public Utility: History, Theory, and Practice

3.1 INTRODUCTION

Telecommunications may be described as a *process* that enables individuals and machines to communicate across space. In order to communicate, a medium or *technical system*, is required. The essential components of the modern, technical system for telecommunications are based on *electronic* and advanced *optical* technology. Access to the system is provided as a commercial *service* by private corporations. In the United States and Canada the operations of these corporations are subject to a particularly invasive and extensive type of state control. This is because the telecommunications industry belongs to a set of infrastructure industries that have historically been identified as natural monopolies and accorded the status of *public utilities*. Other public utilities include the water, gas, and electricity industries.

It follows that there are a number of approaches to the study of telecommunications. One may study telecommunications as a process (e.g., from a socio-psychological approach), as a technical system (as an engineering problem), as a service industry (from an economic perspective), and finally, as a public utility. In this chapter we will study telecommunications by looking at its status as a natural monopoly, subject to a particular type of state regulation known as public utility regulation.

In the previous two chapters we analyzed the Canadian and U.S. industry structures by looking at their early history and by considering factors related to industry structure such as ownership, the extent of vertical and horizontal concentration, regulatory jurisdiction, and regulatory frameworks. In keeping with the descriptive nature of this analysis, we did not consider in detail the justifications for the regulation of public utilities in general and telecommunications in particular. Nor did we analyze the particular form that this regulation has taken in the United States and Canada. Regulation was simply presented as a commonplace of the industry.

41

In this chapter we will consider why the telecommunications industry has been singled out for a particular type of regulatory treatment. In other words, we will consider in more detail the *historical* justifications for regulation and will undertake to examine the *theoretical* justifications for the regulation of common carriers and public utilities. We will consider such basic theoretical questions as: What is a natural monopoly? What are the criteria for identifying natural monopolies? Why do natural monopolies need to be regulated? What are the goals and objectives of this regulation? How effective is this regulation? In order to respond to these questions we will draw on perspectives taken primarily from economics and political theory.

Economic regulation presupposes the existence of an agency responsible for controls on market entry, on prices, and on the quality of service. In this chapter we will see that the modern regulatory agency is a governing instrument of relatively recent invention, one that owes its creation to a historically specific set of circumstances. Moreover, we will see that U.S. and Canadian regulatory agencies have been created during three different historical periods in response to three distinct problematics. In other words, all regulatory agencies are not created the same. In order to understand the structure, the powers, and the limitations of a regulatory commission such as the Federal Communications Commission (FCC) or the Canadian Radio-television and Telecommunications Commission (CRTC), it is essential that it be identified with the period of its inception and the social and political debates that influenced its creation. Finally, we will see that the FCC and the CRTC are essentially hybrid organizations: they are part policy maker, and part adjudicator. In part two we will consider how this duality shaped both the pace and the modalities of U.S. and Canadian deregulation.

3.2 REGULATION AND THE STATE

The modern state has three governing instruments at its disposal: spending, exhortation, and regulation (Doern 1978, 14). Governments may spend in order to meet their commitments to voters. They may attempt to educate or persuade citizens and organizations to adopt desirable behaviors. Finally, governments have the option of regulating the behavior of citizens and corporations.

Spending is the most appealing option for governments. However, there are inherent limits to spending as a governing instrument since it is contingent upon revenues generated by taxes. When spending exceeds revenues generated by taxes, governments have the option of borrowing money, but there are limits to this as well. Taxing and spending effect a redistribution of wealth between different groups of citizens, and between citizens and private corporations. When taxing becomes excessive, the spending option may lose its appeal as different groups in society lobby the government for tax reductions and concessions.

Persuasion in the form of public information campaigns is an appealing governing instrument because costs are usually not recurrent. However, it is difficult to predict the success of information campaigns. The modification of certain types of behaviors may not be susceptible to persuasion. Moreover, it may take many years for this type of governing instrument to meet its objectives.

Regulation is the most heavy-handed of governing instruments. In the modern economy it is difficult to imagine a business that is not affected in some way by government regulation. For example, there are government controls on the labor market, the right to unionize, the length of the workweek, and the requirement for a minimum wage. There are standards for safety in the workplace. There are standards related to the quality of products, for example, in the food industry. There are controls on the use of certain substances, for example, on the use of pesticides in agriculture. There are standards and codes applicable to the construction of buildings. Safety regulations govern the operations of the airline, automobile, and transport industries. Regulation is a fact of life for all industries in the modern economy. It is not, therefore, the fact of government regulation that distinguishes telecommunications from other industries; rather it is the mode of regulation that is exceptional.

Alfred Kahn in his seminal work, *The Economics of Regulation: Principles and Institutions* (1971), notes that most government regulation is designed to supplement the activity of markets. That is to say, it attempts to compensate for the imperfections of the market. For example, government controls on the processing of meat are designed to protect the health of consumers and to create standards for the grading of meat to provide consumers with objective information on the quality of meat products. The presumption underscoring this type of regulation is that a totally free market would provide neither sufficient protection to consumers from the sale of contaminated meat, nor accurate information to consumers on the quality of meat products. While these controls may add to the cost of meat production, *they do not interfere with the essential activity of the market*. They do not interfere with the process through which prices for goods are set under conditions of free commerce between buyers and sellers. They compensate for the imperfections of markets, while leaving the market intact to perform its basic functions.

Public utility regulation, on the other hand, seeks to replace the market. It does not supplement the market; it *supplants* the market. Public utility regulation is not a response to market imperfections; rather, it is a response to conditions of *market failure*. The goal of public utility regulation is not to compensate for market imperfections but to correct for market failure. It does this by substituting government regulation for normal market activity. Under public utility regulation the power to set prices, to control entry into an industry, and to set standards for the quality of service is transferred to a regulatory agency of the state. Under normal conditions, the market performs these functions. By setting prices, controlling the number of firms in an industry (entry), and setting the standards for service quality, the regulatory agency is usurping functions usually performed by the market.

To summarize, we see that regulation is the most heavy-handed and coercive of the three types of governing instruments and that within the regulatory type, public utility regulation is the most extreme manifestation of this form of government control. No doubt there is a political dimension to this, since in a free market economy, public utility regulation is a potent symbol of the limitations of markets. This should be kept in mind as we consider in subsequent chapters the recent history of attempts to deregulate the telecommunications industry by challenging its status as a natural monopoly and by challenging the efficacy of public utility regulation as a governing instrument.

3.3 THE SOCIAL CONTROL OF COMMERCE: A BRIEF HISTORY OF ECONOMIC REGULATION

A basic component of public utility regulation is the social control of prices. The practice of regulating prices set by private industry has a long history. During the first centuries of the Christian Church, the church fathers—following the thought of St. Augustine (354–430)—expounded the doctrine of *justum pretium*, that is, just price. The doctrine challenged the concept of *verum pretium*, that is, natural price, as embodied in Roman law. It did this by drawing attention to the potential for coercion in exchange. According to the early Church, trade was acceptable only in as much as "the trader paid a 'just price' to the producer, and in selling, added only so much to the price as was customarily sufficient for his economic purpose" (Glaeser 1957, 196). The doctrine was meant to discourage unjust enrichment. According to Martin Glaeser, it was the influence of Christian theology that led Emperor Diocletion to issue two price edicts in A.D. 285 and 301.

The concept of just price survived into the Middle Ages when St. Thomas Aquinas (c.1225–1274) reiterated it. Aquinas emphasized that buying and selling should occur to the common advantage of both parties (Glaeser 1957, 196). Exchange should be based on a just price, which was not necessarily the same as the market price. But what factors should go into the calculation of a just price? According to Aquinas a just price takes into consideration the various expenses associated with production. These include "labor and material costs, cost of transport including compensation for danger and risks, and costs of storage" (196–197). Interest was also a legitimate expense, in spite of the Church's condemnation of usury. We see from this that there is a Western tradition of treating exchange as more than a simple economic relation, as *exchange also entails a social relation with ethical dimensions*. According to this tradition, it is not simply a case of "whatever you can get away with." Where exchange occurs under conditions of coercion, society has a legitimate right to impose controls. One of the bases for the social control of economic activity had been established.

Fundamental to public utility regulation is the premise that not all economic activity is equal from a societal point of view. Certain types of economic activity have a public dimension. In the medieval economy, a distinction was made between private and public (or common) employment. Private employment described the case of craftsmen that were in the permanent employment of the feudal lord. Common employment was used to signify labor that was available to all members of society. Thus, medieval law refers to such occupations as the common innkeeper, common tailor, common blacksmith, common barber, and common miller. In the medieval economy these common trades and crafts were organized and controlled by guilds, a form of local monopoly. Underscoring the medieval guild system was the premise that these common or public trades required close regulation. For example, the medieval guild set prices and standards of quality for their goods, and agreed on wages for workers. The only people who could sell goods were those that belonged to a guild. In nineteenth-century jurisprudence, the term "business affected with a public interest" was used to justify the introduction of price regulation of monopolists. Although this is a much more restrictive use of the concept of public or common business than we find in the Middle Ages, the modern concept is derived from common law and the medieval guild system.

The medieval guild operated as a local monopoly. During the mercantile period (from the 1500s to the late 1700s), markets widened and the nation-state emerged as an economic system. It was during the reign of Louis XIV that Jean Baptiste Colbert (1619–1683), Louis's finance minister, developed the economic policy of mercantilism. Colbert favored the intervention of the state in economic activity as a way of enhancing the economic wealth of the nation. According to Colbert, a nation's strength was a function of its balance of trade— usually measured in terms of its reserves of precious metals. Colbert's policy was followed by the other great European nations of the period: England, the Netherlands, Spain, and Portugal. In order to ensure a favorable balance of trade, tariffs were introduced to protect national industries. In order to encourage economic expansion, royal charters were granted to trading and plantation companies. In granting these exclusive charters, the state was underscoring the fact that these enterprises had been established to contribute to the wealth of the nation. In effect, these chartered companies were instruments of the state, created to meet state objectives. By granting these companies exclusive rights, governments hoped to reduce risks to investors, thus spurring investment. Because they operated as monopolies these joint-stock companies were often very profitable. These privately owned companies, operating under monopoly charters, were in some respects the precursors of the public utility companies of the nineteenth and twentieth centuries.

Under the economic policy of laissez-faire of the eighteenth and nineteenth centuries, the pendulum swung the other way. The doctrine of laissez-faire prescribed an end to government interference in the economy. Under the new doctrine, free commerce, unfettered by government interference, was viewed as the best way to stimulate economic growth and by extension to increase the wealth of the nation. Under laissez-faire it might be necessary for governments to intervene to break up monopolies, but certainly not to create them.

We will now consider in greater detail how the U.S. judicial system adapted itself to the requirements of nineteenth-century industrial capitalism. It is important to understand how U.S. common law evolved in the areas of property rights and commercial law during this period, because it was on the basis of this experience that the current economic model, involving state regulation of privately owned corporations, was established. It is largely on the basis of the U.S. experience that a theory of public utility regulation was developed that was applicable to the United States and Canada. Moreover, the evolution of common law in the United States closely parallels that of Canada, whose common law tradition was also derived from England. Although the model is based on private ownership of industry, it is important to remember the numerous ways that private enterprise is supported by the legal system and by a government policy designed to reduce risks and encourage investment in key sectors of the economy. Also fundamental to the development of public utility regulation (as applied to the telecommunications sector) is the common law concept of the *common carrier*. We will see that governments were active in supporting the growth of railroads, a quintessential common carrier industry. When opposition to the power of the railroad monopolies grew, the framework already developed for the regulation of traditional common carriers was applied to the emerging railroad industry. This common carrier framework would eventually be applied to the regulation of the telecommunications industry.

The political culture of the United States, founded as it was in a rebellion against the centralizing authority of the British Crown, was particularly resistant to federal intervention in the business of the nation. The U.S. Constitution provides for the formal separation of powers between the executive (the presidency), the legislative (Congress), and the judicial (courts) branches of government. When local partisan politics prevented the executive and legislative branches of the U.S. government (at the federal level) from promoting interstate commerce, it was the judicial branch that took the lead. The courts did this by adapting the legal system in a way that "guaranteed private entrepreneurs certainty and predictability of economic consequences" (Horwitz 1989, 51).

In specific terms, the courts affected a shift in the common law approach to property rights. In traditional common law, property rights were associated with the rights of the aristocracy to use (enjoy) their property for recreational purposes. If my neighbor's use of his or her land (e.g., the damning or redirecting of water) deprived me of the enjoyment of my property, the courts would generally support my claim against the neighbor. This approach was essentially opposed to economic development. In order to encourage development and create a more stable environment for private industry, the courts began to look more favorably on the *productive use* of property (Horwitz 1989, 51). The courts did this by giving priority to the first entrepreneurs to establish a business. For example, the courts would tend to protect the rights of the first miller to harness a river or stream against later competitors whose businesses further up stream might reduce the flow of water (and hence power) to the first miller. Giving priority to the first entrant was justified because his or her risks were considered to be greater.

The courts would also affect changes in tort law that had the effect of reducing the liability of corporations against claims for damages. Under traditional tort law, compensation was due to a plaintiff who could demonstrate that the accused had caused him or her damage or nuisance. After the 1840s the burden was on the plaintiff to demonstrate that the nuisance or damage was the result of *careless* or *negligent* action. The shift from nuisance to negligence reduced chances for success in a tort action and thus encouraged economic development (Horwitz 1989, 52).

At the level of constitutional law, the U.S. Supreme Court rendered decisions that limited the power of the states to regulate the activities of corporations. The landmark case was *Dartmouth College v. Woodward* (1819). Dartmouth College in New Hampshire was chartered by King George III in 1769. In the early 1800s the state of New Hampshire tried to gain control of the college by appointing members to the board of trustees and altering its faculty. The college resisted these changes and the Supreme Court found in favor of the college. In doing so the Supreme Court sanctioned the transformation of Dartmouth from a corporation with public origins to a private one. This had the effect of recognizing business corporations as essentially private entities. Robert Britt Horwitz notes that this "began the long process of attempting to distinguish between legally public and private corporations" (1989, 55).

By giving preferential treatment to first entrants and the productive use of property, by transforming the criteria for tort actions from nuisance to negligence, and by promoting the independence of private corporations from state governments, the courts were formulating a de facto economic policy that encouraged private ownership and investment. The legislative and executive branches of government were also active in the promotion of economic growth during this period. A major focus of government policy was the transportation sector. During the period of western expansion in Canada and the United States, the development of the transportation infrastructure was seen by governments as essential to the task of nation-building. One of the most important instruments at their disposal was the power to grant exclusive franchises to build canals, roads, and bridges. Another option entailed the use of public funds to support the development of transportation systems. Concerning the United States, Horwitz notes: "Overall, 70 percent of the cost of canal construction from 1815 to 1860 was provided by public investment. Until 1961, government agencies provided close to 30 percent of the entire investment in railroad building, *not* including land. The one major case of direct financial aid from the federal government was a post–Civil War loan of $65 million to the railroad companies that constructed the first transcontinental line" (1989, 53). In chapter two we considered the Canadian government's ownership of Canadian National railways, a form of direct investment in the transportation infrastructure.

The justification for granting franchises, preferential loans, and in the case of Canada, outright public ownership of the railroads was the centrality of transportation to other forms of economic activity. It was not important for nation-building that the government intervene directly and massively in the clothing or agricultural sectors. But it was important that the government create conditions favorable to trade by supporting the development of a transportation infrastructure. But this involvement did not go unopposed. In their efforts to foster transcontinental (interstate and interprovincial) trade, the federal governments of the United States and Canada met with opposition from local and state/provincial authorities. One of the modern legacies of this clash in the area of telecommunications would be jurisdictional conflict between state/provincial and federal governments.

Notwithstanding the case of Canadian National railways, the dominant mode of economic development in both Canada and the United States was based on private ownership. Governments supported private development of the transportation system through loans, land grants, and exclusive franchises. But the private sector model as applied to the railroads, the first truly national industry, led to abuses; these led to the promulgation of populist legislation designed to curtail the power of the railroad companies. Ultimately, it fell to the courts to resolve the conflict between private economic interest and the public interest.

The judicial case that had the most direct bearing on public utility regulation in the United States was the Supreme Court decision in *Munn v. Illinois* (1877). As the railroads expanded into western United States, they became the principal means of transportation for purposes of commerce. As the industry became more powerful, local and regional interests challenged this power by accusing the railroads of unequal treatment and rate discrimination. Anti-monopoly laws were enacted by the western states of the Mississippi Valley (Illinois, Iowa, Minnesota, and Wisconsin) to regulate railroad rates and to regulate the activities of grain elevators. These laws were known as the Granger Laws. The grain warehouse firm of Munn and Scott, which owned a grain elevator in Chicago, set its prices at a level higher than what was permitted by the Illinois act. At issue was the right of the state of Illinois to set rates for the grain elevators. In this case the Supreme Court decided in favor of the state of Illinois. The basis for the decision would be particularly significant for the subsequent evolution of public utility regulation. Glaeser has described the rationale behind the Supreme Court ruling in the following terms: "under the circumstances in which the elevators were being operated in Chicago (that is, standing 'in the very gateway of commerce and taking toll from all who pass,'), they had become a business 'affected with a public interest and had ceased to be *juris privati* only' " (1957, 207). In other words, although the grain elevators were privately owned, their status as monopolists, operating a service that was indispensable to the general commerce in grain, invested the company with a public interest. Society, through the instrument of the state, had a legitimate interest in their activities that extended to the fixing of prices.

How are we to interpret the phrase business "affected with a public interest"? It would appear to be a very broad concept, applicable in varying degrees to any number of businesses. Chief Justice Morrison R. Waite appears to have endorsed a generous interpretation when he stated: "When, therefore, one devotes his property to a use in which the public has an interest, he, in effect, grants to the public an interest in that use, and must submit to be controlled by the public for the common good, to the extent of the interest he has thus created. He may withdraw his grant by discontinuing the use; but so long as he maintains the use, he must submit to the control" (Glaeser 1957, 207).

The *Munn* decision cut several ways. On the one hand, it recognized the rights of states to regulate prices. On the other hand, it restricted this right to cases where there was a manifest public interest *as a result of the presence of a monopoly,* or *near monopoly.* Finally, it was up to the courts to decide whether a business was clothed with a public interest, not to the individual states. As a result of this decision, the association between monopoly and the public interest had been established. Equally significant, the decision recognized the right of government to usurp market price setting where the public interest was clearly at stake.

The *Munn v. Illinois* decision marked the triumph of public interest over private interests. In subsequent years, however, the pendulum would swing in the other direction. The Supreme Court of the United States reversed itself, relieving the railroads of a variety of state regulations. It soon became evident, however, that an uncontrolled railroad industry was not serving the interests of either the railroads or their clients. Excessive competition was damaging to the industry. Railroad bankruptcies were frequent. There were problems financing routes to remote areas with less traffic. In order to compensate for their losses on competitive routes, the railroads would inflate prices on segments where they held a monopoly. When this occurred, branch routes were charged rates that did not reflect costs. Shipping rates were not set on the basis of miles traveled, but rather on the railroad company's ability to profit from its monopoly. In order to introduce some stability to the industry, the railroad companies attempted to form secret cartels. Finally, it became apparent that Congress would have to step in. Congressional action took the form of the Interstate Commerce Act (1887), that created the Interstate Commerce Commission (ICC). Principal among ICC's powers with respect to the railroad industry was to ensure that rates were reasonable and that there was no price discrimination.

In Canada the regulation of railroads began in 1868 when the Railway Committee of the Privy Council was given oversight responsibility for the railroad industry. The Department of Railways and Canals was created in 1879. In 1888 the Railway Act was passed. The law stipulated that railroad rates had to be *reasonable and non-discriminatory*. In 1903 the Railroad Committee of the Privy Council was abolished; the Board of Railway Commissioners, composed of three members, replaced it.

What was the justification for regulating the railroad industry? Although there was competition on heavily traveled routes, many of the routes were served by a single company. From the point of view of the public interest, it was not just that the railroads operated as monopolies (or near monopolies). They were monopolies operating as *common carriers*. The common carrier tradition in common law was an old one. The first common carriers were coachmen, teamsters, ferrymen, and operators of canal boats. As the industrial era progressed, a number of public callings, such as the common tailor, barber, and surgeon would lose their public status. But this was not the case with respect to common carriers, whose operations continued to be subject to public scrutiny. Presumably, this was due to the essential role they played in the commerce of the nation. When the railroad industry emerged as a major mode of transportation, the common law framework of the common carrier was applied to it. Horwitz has described the duties of common callings such as the common carrier in the following way: "they must serve all, they must provide adequate (and safe) facilities, they must charge reasonable rates, they must not discriminate against customers" (1989, 59).

We see that the emergence of the common carrier regulation of the nineteenth and twentieth centuries, while running counter to the ideology of laissez-faire, drew upon and was bolstered by earlier economic traditions and modes of thought justifying the social control of economic activity. These traditions temper the impulse in the industrial period of laissez-faire capitalism towards a total withdrawal of the state from the regulation of economic activity. From the early Christian Church emerges the admonition against coercion in economic relations and the principle that exchange should be fair to both parties. There should be no excessive enrichment. From the Middle Ages is derived the concept of common or public callings. The mercantile period established the precedent of state sponsored monopoly as an instrument of public policy. When the concept of common callings is combined with the mercantile precedent of chartered monopolies, a fundamental precept of public utility regulation emerges: When an industry plays a central and essential role in the nation's general economy, it is the legitimate responsibility of the state to ensure that the industry function efficiently and fairly for the benefit of all of society. This support may take the form of government sanctioned monopoly. The decision in *Munn v. Illinois* established the connection between monopoly and the public interest: The state had a legitimate interest in the prices charged by a monopolist operating at the "gateway of commerce." If necessary, governments could intervene to set maximum prices for these operators.

Our review of the justifications for the social control of economic activity has thus far emphasized the case of the common carrier. In chapter one we considered Theodore Vail's slogan "One system, one policy, universal service," which sanctioned economic regulation of telephone *common carriers*. In many respects, however, the association between telecommunications and common carriage is a tenuous one. Telecommunications has more in common with another industry group comprising the water, electricity, and gas industries. All of these industries share a common characteristic: they depend on a direct connection, a physical link, between their facilities and the consumer in order to provide service. This is also the case with the telephone industry. In actual usage, the term "public utility" has come to designate industries that require a direct link to the consumer. It is customary, however, to use the term "economic regulation," to refer to the mode of regulation of both industry groups. This is the case because, although different, common carrier and public utility industries share a common trait: they are subject to market failure requiring price regulation in order to protect consumers. However, the type of market failure differs. In the case of the railroads, the problem was excessive competition in long-haul markets accompanied by no competition (monopoly) on short-haul routes. In the case of direct-link industries, market failure is related to characteristics of the technology of production that make it difficult, if not impossible, for the industry to sustain efficient competition. Stated in other terms, this means that one firm can supply the service or product more efficiently, and at less expense to the consumer, than two. When such a condition exists, the industry is said to exhibit the characteristics of a natural monopoly. But there is no definitive, universally acknowledged test for natural monopoly. Nevertheless, most scholars would agree that the railroads were never natural monopolies. To summarize, the two "common carrier" industries (railroad and telephone) did not share a common problem, but they did share a common solution: the regulation of prices and the control of entry.

3.4 THE BIRTH OF THE MODERN REGULATORY AGENCY: THE THREE PHASES OF ECONOMIC REGULATION

In the late nineteenth and early twentieth centuries a new type of governing institution, the regulatory agency, was created. The ICC, whose creation was described in chapter one, is an example of one of the first regulatory agencies. It would be followed by numerous other regulatory bodies, including the FCC and the CRTC. By the mid-1960s a concerted critique of the performance of these regulatory bodies had begun to coalesce and attract the attention of policy makers. In chapter four we will review this attack on the regulatory agencies. However, in order to fully understand this critique, a basic understanding of the origins of these institutions and their relationship to forms of political authority (government) is required. In this section and the coming sections we will consider these dimensions of the modern regulatory agency.

In *The Irony of Regulatory Reform* (1989), Horwitz identifies three phases of economic regulation in the United States: the Progressive Era (1900–1916), the New Deal Era (1930–1938), and the Great Society Era (1965–1977). During each era, the U.S. government intervened to resolve a specific set of problems that were endemic to the economy of the period (table 3.1). These periods are also relevant to an understanding of the circumstances that gave rise to economic regulation in Canada.

During the Progressive period (1900–1916), governments intervened to impose national standards and regulation. These regulations were not sectorial, that is to say, they were not specific to a particular industry. The preoccupation of regulation during the Progressive Era was the problem of economic concentration: how to deal with concentrations of economic power that throttled competition and the functioning of markets. The Sherman Antitrust Act of 1890 made "every contract, combination in the form of trust or otherwise, or conspiracy, in restraint of trade or commerce . . ." illegal (Galbraith 1987, 161). During the Progressive Era an attempt was made to further define activities that fell under antitrust law. The agencies created during this period included: the ICC, the Federal Reserve System, and the Federal Trade Commission. Their mandate was to ensure fair trading practices over a *national trading area*, with a particular focus on conspiracies in restraint of trade.

The reforms associated with the New Deal period (1930–1938) were not so much a response to concentrations of market power as they were an attempt to deal with *market collapse*. These reforms were directed at particular sectors of the economy and gave rise to regulatory agencies with powers over a single industry. Some of the agencies created during this period were the following: the Federal Power Commission (with the power to set prices for interstate electricity); the Food and Drug Administration; the Federal Deposit Insurance Corporation (with responsibility for supervising insured banks); and the Civil Aeronautics Board (with responsibility for the airline industry). During the New Deal period the thrust of regulation was to rescue industries that were on the verge of

Table 3.1   Three Periods of Economic Regulation

Dates	Period	Sector	Examples	Objectives
1900–1916	Progressive	No	Sherman Act	Control conspiracies in restraint of trade
1930–1938	New Deal	Yes	FCC	Reestablish economic stability during periods of crisis related to overproduction
1965–1977	Great Society	Yes	EPA	Reverse the consequences of negative externalities

collapse. Often this was a result of too much competition during a period of economic depression. The regulatory response was to introduce price stability and controls on production so that these industries could become profitable again. In effect, these New Deal agencies were established to create economic cartels and supervise their activities (Horwitz 1989, 69).

The FCC, created in 1934 with authority to regulate the broadcast, telephone, and telegraph industries, was a case in point. Without some mechanism for allocating radio frequencies, broadcasters were interfering with each other's signals. The FCC's role was to bring order to the market by controlling entry through its power to grant licenses. This had the effect of creating a cartel of commercial radio networks. By the time authority to regulate the telephone industry had been transferred from the ICC to the FCC, AT&T already exercised a monopoly. The Communications Act of 1934, which created the FCC, merely lent legitimacy to this control. A common attribute of the New Deal agencies was the setting of prices and the control of entry in a given industrial sector. Although these agencies served the interests of consumers through their powers to control prices, they also served the interests of the industries they regulated by controlling competition. The consensus that existed between government and telephone industry during this period has been summarized by Michael K. Kellog, John Thorne, and Peter W. Huber in the following way:

> For quite different reasons, many players in the industry came to favor federal regulation. The independents wanted it to stop Bell and obtain interconnection. State and local governments hoped to end "inefficient competition" and expand service. Bell itself wanted to consolidate its dominant position and legitimize a monopoly— its own. Bell's chairman, Theodore Vail, spoke publicly in favor of regulation, sounding a theme ("cream skimming") that would become a Bell System rallying cry for the next half century. Despite its own record of having sought out the most profitable routes, Bell now resolved "to serve the whole community" and embraced regulation to avoid "aggressive competition" covering only that part that was profitable. (1992, 79)

During the Great Society Era (1965–1977), the impetus for regulation was the control of "the social consequences of business behavior" (Horwitz 1989, 76). In the language of economics this meant the control of *externalities*. Examples of regulatory agencies created during this period include the Environmental Protection Agency (1970), the Consumer Product Safety Commission (1972), Occupation Safety and Health Administration (1973), and the National Highway Traffic and Safety Administration (1970). The principal orientation of these agencies was environmental, occupational, and consumer protection (77).

With the possible exception of the Progressive Era, the introduction of economic regulation to Canada followed a pattern similar to that of the United States. As was the case in the United States, it was imperative that the Canadian economy of the late nineteenth and early twentieth centuries make the adjustment from local/regional spheres of economic activity to a national trading arena. But there was less enthusiasm in Canada for antitrust measures, such as the U.S. Sherman Act. In order to promote national industries, the provinces and the federal government often worked in partnership with industry. A 1981 report by the Economic Council of Canada described this relationship in the following manner:

> In the early years of Canada's history, the role of government was not so much to restrain individual activities as to support initiatives that would help open up the country and develop its resource and industrial potential. The understanding that government was, in a sense, in partnership with private industry in this development venture led as well to favorable relations between the federal government and the dominant economic interests of the period. (Economic Council of Canada 1981, 2)

Two examples of this partnership relationship were high import tariffs, which favored eastern manufacturers, and agricultural support programs, which favored the western provinces.

During the years of the Great Depression, regulatory initiatives followed a pattern similar to that of the United States. The problem of excessive competition in a period of economic decline was met with legislation primarily in the agricultural and transportation sectors (rail and road transport):

> The difficult economic conditions of this period fostered doubts about the advantages of market mechanisms and gave rise to new demands for policies providing special protection. Low wheat prices all but destroyed the wheat pools created in the 1920s by farmer-controlled co-operatives and induced the federal government to reconsider the grain growers' demand for a government marketing agency. . . . Trucking was another industry that was particularly hard hit by the economic conditions of this period. . . . The appeal of motor carriers for entry controls, which was strongly supported by railroad interests concerned about the "unfair competition" that rail was facing from the highway mode, was met by legislation governing entry in a number of provinces. (Economic Council of Canada 1981, 3)

From the late 1960s through the 1970s, Canadian legislation was enacted in the areas of worker, consumer, and environmental protection. Some examples include the Hazardous Products Act (1969); important revisions to the Food and Drugs Act (1963); the Motor Vehicle Safety Act (1970); the Consumer Packaging and Labeling Act (1974); and the Clean Air Act (1971). This followed the pattern of legislation dealing with negative externalities in the United States.

Albeit brief, this review of regulatory periods suggests that there was not a single impetus for the introduction of economic regulation. Regulation has been introduced in the twentieth century as a response to varied economic conditions, corresponding to distinct and different types of economic problems: concentration and collusion by producers, excessive competition in periods of economic decline, and as a response to externalities. Moreover, the introduction of economic regulation was not solely, or even primarily, driven by consumer interests. Particularly in the period of the Great Depression, it was supported by producers who favored controls on prices and entry as a way of introducing greater stability to their industries in a period of economic crisis. The introduction of regulation to infrastructure industries such as railroads and telecommunications belongs to this period.

3.5 SOME PRINCIPLES AND PRACTICES UNDERSCORING THE ECONOMIC REGULATION OF TELEPHONY AS A PUBLIC UTILITY

Once the telephone industry accepted government control in exchange for protection from competition, it became the job of the regulators to determine the best way of exercising this control. An essential component of this control was the power to set rates for the different types of telephone service. This was no easy task since in order to set rates in a logical and consistent fashion standards and objectives for rate setting had to be agreed upon. A simple solution would have been to respond to telephone company requests for higher rates by reducing them by a set percentage, for example, by 50 percent. Should the telephone company request a dollar a month increase for local service, the regulator could respond by acquiescing to a fifty-cent increase. The disadvantages of this approach are clear, since there is no guarantee in such a negotiated response that a fifty-cent increase would be either necessary or sufficient to ensure the efficient provision of telephone service.

The decision to raise or lower rates had to be based on some objective criteria; if not, the regulator ran the risk of either gouging consumers or crippling the telephone company. A convenient response to this problem was to suggest that rates should be set at a level equivalent to the market price if market conditions were to exist. Of course, there is no way of knowing what the market price would be, since the market had been usurped by regulation. Such an attempt to hypothesize and theorize about what the market price would be under competitive conditions was preferable to blind negotiation, but like all theoretical approaches it would depend on the appropriateness of the models used to simulate competitive conditions and on the accuracy of the information that was fed into the models. To illustrate the complexity of the problem we will describe the theoretical case of monopoly pricing under conditions where the monopolist is *not* subject to regulation. This case is often used to explain in economic terms the likely behavior of the unregulated monopolist, and why such behavior would not be optimal from a social point of view. In the following scenario the term "widget" is used to designate a non-specific, imaginary unit of production.

Let us assume that a company can manufacture a widget for four cents and that the cost of producing each new widget will not increase or decrease with changes in the volume of production.[1] It will cost the same amount to produce widget number 50,000 as it will to produce widget number 150,000. In other words, there are no economies of scale. On the consumption side let us assume that the demand for widgets is sensitive to price. As the price goes up, the number of consumers disposed to purchase the widget declines. As the price goes down, the demand for widgets increases. Assume that the relationship between price and demand is such that at a price of seven cents the monopolist is able to sell ten thousand widgets, at a price of six cents the monopolist is a able to sell twelve thousand widgets, at a price of five cents thirteen thousand widgets, and at a price of four cents fourteen thousand (table 3.2). When the monopolist sets the price at four cents, his price is equivalent to what economists refer to as his *marginal cost*. A simple calculation tells us that given this price-demand equation for widgets, the monopolist will derive greatest profit *by selling fewer widgets at a higher price*. As paradoxical as this may seem, a comparison of profits derived at each of the price-demand levels bears this out. For the purposes of comparison we will consider the case of widgets sold at seven cents and five cents (table 3.3). The sale of ten thousand widgets at seven cents will produce revenues of $700, at a cost of $400 (ten thousand × four cents), generating a profit of $300. The sale of thirteen thousand widgets at five cents will produce revenues of $650, at a cost of $520 (thirteen thousand × five cents), generating a profit of $130. The profits generated at seven cents will be greater than those at five cents in spite of the fact that three thousand fewer widgets have been sold. Under these conditions there are clear incentives for the monopolist to sell fewer widgets at a higher price.

Table 3.2   Demand Elasticity for Widgets

Price	Consumer Demand
4 cents	14,000
5 cents	13,000
6 cents	12,000
7 cents	10,000

Table 3.3   Increasing Profits by Selling Fewer Widgets at a Higher Price

Production Costs	=	4 cents × 13,000 = $520
		4 cents × 10,000 = $400
Revenues	=	5 cents × 13,000 = $650
		7 cents × 10,000 = $700
Profits	=	$650 − $520 = **$130**
		$700 − $400 = **$300**

Economists refer to the price-demand relationship as market, or price elasticity. When prices are *inelastic*, changes in price do not effect demand. When prices are *elastic*, demand is sensitive to changes in price. Given the market elasticity for widgets described above, what is wrong with the monopolist selling fewer widgets at a higher price in order to maximize profits? Stated simply, the problem is that there are four thousand potential consumers who are prepared to meet the widget maker's marginal costs of production (i.e., four cents) who are deprived of their use. Moreover, this situation can only exist if there is no competition. If the industry were subject to competition, another firm could be expected to enter the market and meet the residual demand for widgets. Fierce competition among several firms would result in prices close to marginal costs. From a societal point of view, the problem is not so much that the monopolist's prices are excessive, but that by pricing in such a manner, a large number of consumers who are able to offer adequate compensation to the widget-maker are denied the potential benefit of widget use. Moreover, the situation is artificial, since it is only the absence of competition that makes this possible.

We see that the problem of *monopoly pricing* is not simply a case of excessive profits or of price gouging. When a monopolist charges monopoly prices, he or she is depriving society of the collective benefit of the use of widgets up to a level of demand that is capable of meeting marginal costs. When the problem is framed in this way, we see that the problem of monopoly is not simply one of unfair prices. More importantly, it is a problem of what economists refer to as *allocative efficiency*. When the monopolist engages in monopoly price setting, society's resources are not being allocated efficiently. According to economic theory, competition is the best mechanism for allocating resources since it will have the effect of driving prices down to the level of marginal costs.

What does all this mean for the regulator? At first glance it suggests that the goal of the regulator should be to set rates at a level close to marginal costs (with an allowance for some profit taking by the monopolist). Of course, this assumes that the regulator has sufficient information to determine the monopolist's marginal costs. According to this model of rate setting, based on marginal costs, approximately fourteen thousand consumers would benefit. But what if the regulator, cognizant of the price elasticities for widgets, were to decide to *price discriminate*? In other words, if the regulator, in conjunction with the monopolist, were to create different categories of services and to charge more where the consumer is willing to pay more, and less where the consumer is unable to pay. The case of different rates for business and residential telephone service springs to mind. One might expect that telephone service will be more valuable to the business user than to the average residential consumer. By setting rates higher for basic business service, based on the value of this service to this class of user, it should be possible to lower rates to other classes of consumers such as residential subscribers. In fact, it may be possible to provide service to some users at a price below marginal cost while still allowing for profits. In the case of our hypothetical market for widgets, the total number of subscribers would exceed fourteen thousand.

Although there would be no incentive for an *unregulated* monopolist to provide service below marginal cost, it is reasonable to assume that some form of price discrimination above marginal cost would be profitable. Profits would be greatest for the unregulated monopolist under conditions where he or she could create several price tiers based on demand elasticities. Such a pricing strategy would work because there would be no competition to drive prices down to marginal costs. The incentives for the monopolist to engage in price discrimination are clear. But what would be the justification for a regulatory agency, which does not have a mandate to maximize profits, to mimic this behavior? Closer analysis reveals that there may be *economic* reasons to price discriminate in order to increase the availability of a service to consumers. In the case of the telephone industry, the economic justification for extending the availability of the service is based on the presence of what economists refer to as *network externalities*.

Externalities are consequences of economic activity *that are not accounted for in the price and market system*. For example, a company producing electricity from coal may be able to provide cheap electricity to its consumers. But the price it charges does not take into account the true *social* cost of production. Clearly, if the electric company creates pollution (e.g., in the form of acid rain) that kills lakes and trees hundreds of miles away, one cannot say that these costs are accounted for in the price paid by local consumers. These costs are born by the maple syrup and tourist industries of the affected regions. The medical costs of pollution are absorbed by medicare and other forms of medical insurance. The case of pollution costs that are not reflected in the price of electricity is an example of a *negative* externality. But there are also *positive* externalities. An example of a positive externality is to be found in the case of communication networks.

In a communications network, such as the telephone network, the value of the network to all subscribers increases as the number of subscribers increases. A network that enables a caller to reach millions of potential subscribers is infinitely more useful and, therefore, valuable than one that reaches only ten subscribers. But this characteristic of the network may not play a part in the *individual user's* decision to subscribe, or more importantly not to subscribe, to the service. In paraphrasing J. H. Rohlfs (1974), Gerald Faulhaber has described this phenomenon in the following manner:

> suppose our hypothetical subscriber would be willing to pay only $12.00 per month to be on the telephone network, but, collectively, the subscriber's friends, relatives, and, possibly, creditors value his or her availability for calling at $9.00. If the telephone company charges $15.00 per month for subscription, our hypothetical customer would not join, even though his or her personal benefit plus the benefit to potential fellow-communicators is worth $21.00. This potential subscriber is failing

to take into account all the social benefits to be derived from his or her subscription. Because some of these benefits are external to the decision maker, economists refer to them as "externalities," and this particular demand interdependence is called the "network externality." (1987, 110)

The previously depicted situation may be described as the *economic* justification for universal service. We usually think of universal service as a social or political goal—a response to public demand for affordable telephone service. The presence of *positive* externalities associated with increased participation in networks suggests that in addition to socio-political justifications for universal service, there are also economic ones. Given these positive externalities, it makes economic sense to subsidize subscribers who otherwise might not be able to afford telephone service. This subsidy would be financed by setting prices based on the demand elasticities for different types of telephone service. Subscribers who place a high value on telephone service would subsidize subscribers who otherwise might not be able to afford it. The subsidy is justifiable on the grounds that the economic transfer from one class of subscribers to another will enable the regulator to compensate for imperfections in a pricing system that does not take into account the collective benefit to *all* subscribers of maximizing the number of telephone users. In the case of the telephone network, this historical subsidy was based on a transfer from long distance to local service. Eventually, rates for long distance service would be set at a level substantially higher than costs. This generated a surplus that was then channeled to local telephone service, where it was used to promote universality.

In order to promote the goal of universal service, regulators adopted two basic pricing principles: *value-of-service pricing*, and *system-wide price averaging*. Value-of-service pricing is the pricing formula that was described in the previous example. It attempts to achieve optimal pricing by taking into account the demand elasticities for different types of services. As we have seen, it is really nothing more than price discrimination in the service of a lofty goal: universal service. The best example of value-of-service pricing is the price structure for business and residential service, where business rates for a single telephone line are markedly higher than residential rates.

System-wide price averaging is another pricing practice based on a departure from true costs. Under this practice, similar rates are charged for similar services regardless of actual costs that may vary. For example, it costs more to install and maintain telephone lines that are far from a telephone exchange than it does ones that are next door to an exchange. However, both groups of customers pay the same basic rate, irrespective of costs. Similarly, the costs of operating a local exchange in an urban area are less than those associated with operating a comparable service in a rural area where distances between the central office and users are greater and numbers of subscribers are fewer. Under system-wide price averaging, urban and rural consumers are charged the same rate for equivalent service. This rate is based on the average cost of providing service to subscribers (urban and rural) throughout the system. A final example would be that of pricing for long distance. The price of a long distance call is based on the distance between calling points. But the *formula* used to set the rate is based on a calculation that takes into account the average cost-per-mile throughout the entire system. Although the actual cost of providing circuits on individual routes of equivalent length may vary, these are not reflected in the price to consumers.

The two pricing principles do not ignore costs. Costs are still reflected in the price of service. But the way in which prices are related to costs is calculated in a way that discourages the segmentation of costs and takes into account the ability of different classes of consumers to pay. In the hands of regulators, value-of-service and system-wide price averaging became very effective instruments in promoting the goal of universal telephone service. In 1988 the level of telephone penetration in Canada was above 98 percent (Communications Canada 1988, 21), while that of the United States was above 93 percent (National Telecommunications and Information Administration 1988, 79). Penetration is a measure of the percentage of households that have telephone service.

The traditional form of public utility regulation is usually referred to as *rate-of-return, rate-base regulation*. This reflects the fact that the mandate of the regulatory agency is not simply to promote universal service. There is a further requirement that rates be set in a way that will not only enable the utility company to meet its costs, but will also allow it to turn a profit. Stated in other terms, the regulator must also take into account the *revenue requirements* of the firm. In calculating revenue requirements the regulator considers the company's operating expenses, including the costs of labor and the cost of depreciated capital (i.e., investment in equipment). To this calculation is added an allowable rate-of-return, which typically is a percentage of the firm's net assets. In simplified form the formula looks like the following: $R = E + r \times B$. (R = revenue; E = expenses; B = the rate base [or total net assets]; r = the percentage of return on the rate base.) The percentage of return r is usually set somewhere between 8 and 12 percent. One of the goals of regulation is to ensure that rates are set in such a way that total revenues equal R. In this way the public utility is able to cover its costs and make a profit. In the event that a public utility's revenues exceed R, the company may be required to lower its rates and restore the surplus to consumers (Babe 1990, 171).

To summarize, we see that the fundamental challenge facing the regulation of telephony and other public utilities was how to balance the goal of affordable, universal service with the requirement that the monopolist be allowed to receive a reasonable return (profit) on investment. By implementing value-of-service pricing, cost-averaging, and the rate-of-return revenue formula, regulation was able to meet these dual objectives. Clearly, there were social policy justifications for imposing monopoly regulation. But there were also economic reasons related to the presence of network externalities. Underscoring the practice of public utility regulation was the premise that the goal of regulation was essentially to mimic competitive conditions. In other words, one could say that the goal of regulation should be to produce competitive results under monopoly conditions. This was embodied in the prescription that rates for services should reflect marginal costs to the greatest extent possible. As well, monopoly profits should not depart from long-term, average profit levels for a comparable industry operating under competitive conditions. Where prices did not reflect costs, this could be justified as a price optimizing practice. Moreover, one would expect an unregulated firm with market power to engage in a similar form of price discrimination.

3.6 THE TWO MODES OF OPERATION OF THE REGULATORY AGENCY: ADMINISTRATIVE-ADJUDICATIVE VERSUS POLICY MAKING

In chapters one and two we reviewed the statutory powers of the FCC and CRTC. In this section we will consider a fundamental tension that defines the actual mode of operation of the two agencies. This tension results from the fact that the regulatory agency is part policy maker and part arbiter of disputes. This operational duality is the product of the deficiencies of the legislative acts that created the agencies and stipulated their mandates.

When the first regulatory commissions were created at the turn of the century, their mandates generally did not encompass responsibility for making sweeping industrial policy. Their principal function was to act as tribunals, arbitrating punctual disputes between competing producers, and between producers and consumers. As such, these agencies operated on an ad hoc, case-by-case basis. This was the typical mode of functioning of the adjudicative tribunal. There was little room for comprehensive, forward-looking policy making. In general, these agencies were equipped with legislative mandates whose language was terse, vague, and often ambiguous. For example, in the case of Canada, the Railway Act[2] stipulated that all telegraph and telephone tolls "shall be just and reasonable" (340. (1)), and that a company "shall not in respect of tolls or any services or facilities provided by the company, make any unjust discrimination against any person or company or make or give any undue or unreasonable preference or advantage to any particular person or description of traffic"(340. (2)).

Essentially, the only legislative guidance the CRTC (and the other regulatory boards that preceded it) received as to the intent of Parliament was the requirement that tolls be just and reasonable and that common carriers treat all customers and traffic in a non-discriminatory way. These are essentially subjective standards. Moreover, the commission had to evaluate what was just and reasonable in the context of a monopoly industry structure, where there was no market price to serve as a benchmark. For example, the commission could have linked the price of telephone service to the inflation rate. But it did not do so. As a result, the price of telephone service remained stable as the price for other goods and services rose. Of course, by embracing a policy of low prices the commission was supporting the goal of universal service. This was a laudable objective, widely supported by consumers and governments. From a legal or public policy perspective, however, it was not an objective that had been sanctioned in any legislation. Also, and no doubt more significant, the legislation guiding the commission was mute on the question of industry structure (i.e., on the fundamental question of monopoly vs. competition).

The legislative mandate of the FCC was somewhat more explicit. The Communications Act of 1934 stated the following: "For the purpose of regulating interstate and foreign commerce in communication by wire and radio so as to make available, so far as possible, to all the people of the United States a rapid, efficient, Nation-wide, and world-wide wire and radio communication service with adequate facilities at reasonable charges . . . there is created a commission to be known as the 'Federal Communications Commission' " (Brenner 1992, 255 (article 151 of the act)).[3] Although the mandate of the FCC, as stipulated in the Communications Act of 1934, is more explicit than its Canadian counterpart, it is still vague. For example, there is no mention of universal service. Rather, the commission is to be guided by the prudent goal of making service available "so far as possible, to all the people of the United States" (article 151). Similarly, the statute is cautious and vague with respect to facilities and rates. The provision of service should be based on "adequate" facilities. Charges must be "reasonable." The statute is somewhat more precise with respect to the quality of service provided. It must be "rapid, efficient, Nation-wide and world-wide" (article 151). Again, these are very general criteria for rule making; they leave a considerable degree of discretion to the regulator. On the key question of whether the industry should be structured on the basis of monopoly or markets, the legislation is mute.

We see that very little guidance was provided to both the CRTC and the FCC in the exercise of their function as administrative tribunals. These agencies had considerable discretion to determine rates, and to set standards for the availability and quality of service. The legislative mandates were adequate to case-by-case rule making during the period of uncontested monopoly. For example, the basic common law framework for common carriers, which prescribed that rates be reasonable and service non-discriminatory, was an adequate basis for two-way negotiations with the telephone companies, during the period of monopoly. However, it provided little guidance to regulators on how to deal with their negotiating partners, the monopoly telephone companies, when potential competitors sought interconnection with their networks. Not only did the legislation lack specificity with respect to general policy, it also lacked clarity on how and to what extent the regulatory commissions could step out of the adjudicative role to formulate their own policy on industry structure.

It is important not to overstate the opposition between adjudication and policy making. Hudson Janisch in an article entitled "Policy Making in Regulation" points out that the difference is essentially one of degree. Neverthe-less, he notes, this difference "is of the greatest significance" (1979, 50). Janisch has described the two modes in the following manner:

> In an adjudicative role, an agency concentrates on deciding matters on a case-by-case basis, seeking to apply known principles to the facts of each case. In an extreme form, this leaves little room for policy-making, because it is assumed that the policy is contained in the empowering legislation and the agency simply has to apply it in the particular case. At the other end of the spectrum is the role of conscious policy-maker, wherein a regulatory agency embarks on an essentially legislative role and seeks to set out in advance the factors that it will take into account in deciding individual cases. (50)

When operating in a policy-making role, the regulatory agency establishes in advance the rules and the criteria it will apply to future cases. From the point of view of industry, this approach has the advantage of creating a more predictable environment for strategic planning. From the point of view of the regulator, the policy-making approach is generally presumed to be more efficient. Moreover, during periods marked by an increase in the number of cases facing adjudica-tion, the agency may not have the resources to deal with all of them expedi-tiously. This will result in *regulatory lag*: a situation characterized by systemic, chronic delays in regulatory decision making. Regulatory lag will slow the pace of change, a state of affairs that generally favors the interests of incumbent firms. Indeed, dominant, incumbent firms will often engage in adjudication in the knowledge that although they may eventually lose their case, they can count on regulatory lag to discourage potential competitors.

To summarize, the regulatory agency, whether Canadian or American, em-bodies a fundamental tension between administrative-adjudicative and policy-making roles. When these agencies were established, they were designed to function as administrative tribunals that would operate at arm's-length from gov-ernment. Their legislative mandates equipped them to function on an ad hoc, case-by-case basis. Although the cumulative effect of case-by-case decision making was often the creation of de facto policy, the agencies would open them-selves to controversy and criticism whenever they attempted to engage in overt, comprehensive policy making. Moreover, this was precisely what was required as pressures mounted to introduce competition to regulated industries.

When liberalization occurred, it was a slow process resulting from an accu-mulation of individual regulatory decisions. There was no grand plan. Clearly, this process was no way to orchestrate a monumental reorganization of the telecommunications industry. While proponents of liberalization argued that the regulatory agencies were unresponsive to the needs of business users, op-ponents of liberalization contended that they were out of control. Govern-ments, while generally favorable to deregulation, were critical of agencies when their independence resulted in decisions that ran counter to the inter-

Table 3.4   Powers and Autonomy of U.S. and Canadian Regulatory Bodies

Types of Powers	*FCC*	*CRTC*
Nomination of commissioners	Executive (president)	Executive (federal cabinet)
Nomination subject to approval by legislature	Yes	No
Executive branch can reverse decisions	No	Yes (but only in certain cases)
Judicial branch can reverse decisions	Yes (common practice)	Yes (in theory but rarely in practice)

ests of important constituencies. In both Canada and the United States, the focal point of debate, negotiation, and litigation became the jurisdictional arena opposing state/federal (provincial/federal) levels of government (table 3.4). The provinces and the states were often the strongest voices questioning whether the regulatory boards had a mandate to liberalize and challenged their respective federal governments to make these agencies accountable to elected authority. This was the case because it was the state (and in some cases provincial) level of government that would, ultimately, bear the burden of unpopular rate increases. Clearly, what was required was that the regulatory agencies be given a legislative mandate to liberalize, or alternatively, the explicit authority to formulate policy after consultation with interested parties. Eventually, both the CRTC and the FCC would receive a legislative mandate to liberalize. But this would only occur in the mid-1990s after deregulation and liberalization were so far advanced that they were virtually irreversible. During the seminal period of incremental liberalization and deregulation, the regulatory agencies operated in a political grey zone that was not of their own making. Rather it was a function of their paradoxical status as structurally "independent" agencies, with moot authority to make policy, and a dearth of legislative guidelines to assist them in their role as administrative tribunals.

3.7 THE PROBLEM OF AGENCY INDEPENDENCE: CONTRASTING U.S. AND CANADIAN MODELS

It is common to refer to the CRTC and the FCC as *independent* regulatory agencies. At the time these agencies (and their regulatory predecessors) were created, it was generally believed that an enlightened committee of experts, whose authority was grounded in knowledge and not political affiliation, would be best equipped to arbitrate disputes among producers, and between producers and consumers (Cutler and Johnson 1975, 1402). This was an essentially *apolitical* view of policy making. According to this view, formulating industrial policy was essentially a management problem: An agency comprised of independent specialists would be able to make objective decisions, untainted by political influence or pressures from special interests. In retrospect, it is clear that it was naive to suppose that complex economic problems, involving consumers and numerous industrial actors, could be reduced to problems of management or expertise. Nevertheless, the agencies did exhibit a degree of independence from elected (political) authority. Although the Canadian and American regulatory regimes are similar in many ways, they differ with respect to the degree of independence they accorded to the agency. Indeed, this is one of the notable contrasts between the two systems. As we shall see, the term "independent regulatory agency" may be better suited as a description of the American case.

The U.S. system of government, as instituted in the U.S. Constitution, is based on the separation of executive, legislative, and judicial powers. The president appoints the five FCC commissioners, but their appointment is subject to approval by Congress. Although the executive branch has the power to appoint commissioners, it does not have the power to reverse decisions of the commission or send them back for further consideration. In a similar fashion, Congress does not have the power to rescind decisions taken by the commission. In extreme cases, however, Congress may reverse a decision of the commission by enacting corrective legislation. It is more common, however, for Congress to influence the commission indirectly through threats to enact legislation or through its authority to vote the commission's budget every year.

Another vehicle for the expression of congressional will is the congressional hearing. Congress may call commissioners to testify at hearings where they are required to explain and justify agency policy. The Subcommittee on Telecommunications and Finance of the House of Representatives is one example of a committee that has been active throughout the 1980s and 1990s in scrutinizing FCC efforts to liberalize the telecommunications industry. Finally, the courts have the power to rescind FCC decisions. This has occurred on a number of occasions with significant results. For example, competition was introduced to the long distance market in the United States as the result of a court ruling (*see* the *Execunet* decision, chapter five, section six).

Although at first glance the CRTC may appear to function as an independent agency, it would be constitutionally inaccurate to overstate the commission's independence from the federal cabinet and Parliament. As is the case in the United States, the members of a federal regulatory commission in Canada are appointed by the executive branch, that is, the cabinet. However, the terms of appointment are generally longer, candidates are often drawn from the public service, and reappointments are the general rule. In the United States, it is generally expected that when the executive branch changes political parties, the president will appoint commissioners that reflect the policies of his party. In this respect the Canadian tradition of appointing members to regulatory tribunals may be said to favor independence. On the other hand, as we shall demonstrate, the structural relationship between a regulatory agency and the elected government in Canada is not based, as it is in the United States, on the separation of powers.

As Andrew Roman has noted, under the parliamentary system all power devolves, in theory, from the Crown; and the government is viewed as a "unified whole" (1978, 70). In the parliamentary form of government, the judiciary is not equal to parliament; rather it is created by Parliament. In theory, at least, Parliament has the power to dissolve the Supreme Court by repealing the Supreme Court Act.

The more unified character of the parliamentary system in Canada has implications for the status and functioning of regulatory agencies. Roman notes: "When it comes to the creation of federal tribunals, the unitary concept of 'the Crown' prevails, and there is no attempt to create tribunals which are as independent as possible of the other branches of government" (1978, 70).

In Canada the decisions of regulatory agencies are subject to judicial review and to ministerial or cabinet review. An appeal to the courts, however, is generally limited to errors of law or jurisdiction. For example, the courts may rule on whether an agency has overstepped its statutory powers. In Canada the merits or substance of a decision are potentially reviewable by the minister concerned or by cabinet. In the United States there is no recourse to review of independent agency decisions by the president or Congress. Although the Canadian system allows for government review, this power is not frequently exercised. An example of cabinet intervention was the decision by the Pierre E. Trudeau government in 1973 to reverse a Canadian Transport Commission (CTC) decision granting a rate increase to Bell. (In 1976 responsibility for telecommunications was transferred from the CTC to the CRTC.) In reversing the CTC decision, the cabinet argued that the commission "had failed to take into account the 'socio-economic impact' of the rate increase" (Roman 1978, 74). This could not have occurred under the U.S. system. Finally, the government has the power to issue binding directives to regulatory tribunals. In its 1981 report entitled, *Reforming Regulation,* the Economic Council of Canada summarized its findings on the relative autonomy of the regulatory agency in Canada in the following way: "Regulatory agencies in this country are, as a rule, much less autonomous than their U.S. counterparts, and the depiction of regulatory bodies as distinct entities operating independently of the political system is much less appropriate in Canada" (1981, 4).

3.8 CONCLUSION

We see that the telephone industry has evolved in singular relationship to the state. This relationship has been characterized by both an unusual degree of state intervention in the industry and a high degree of industry dependance on the state throughout most of its history. In order to fully understand the modalities and the legitimations of the economic regulation of the telephone industry, this relationship must be placed in the context of a long history of the social control of economic activity encompassing ethical, legislative, and judicial traditions. It is clear that the modern period of nation-building under conditions of industrial capitalism has drawn from and built upon these traditions. Although the regulation of public utilities is an extreme form of state intervention in the economy, this mode of social control of commerce has been erected on well-established foundations.

The regulation of the telephone industry as a public utility owes itself to numerous factors. It would be inaccurate to assert that regulation was simply a response to conditions of natural monopoly related to economies of scale in production. It is true that telephony entails a direct connection with the consumer. This is a feature that the industry shares with other public utilities such as the water, gas, and electricity industries. The high capital costs associated with providing this type of direct connection may act as a barrier to entry and would appear to be a strong indicator of the presence of natural monopoly, since it is difficult to imagine how multiple connections could be offered on a profitable basis. But as we have seen, this characteristic of telephone technology only explains the local network monopoly. It does not explain the long distance monopoly; nor does it explain the monopoly that existed for the provision of subscriber equipment. Furthermore, it does not explain the concentration of ownership of local exchange companies. During the monopoly period, the telephone companies argued that these components (long distance, subscriber equipment, and local service ownership) of their monopolies were a function of economies of scale and scope. There may have been some truth to this argument. But it is also true that governments in both the United States and Canada did little to protect the independents from unfair competition. One thing is clear: in both the United States and Canada, the government and the dominant telephone companies reached a deal that served both their interests. The telephone industry received protection from competition. Universal, affordable telephone service was popular with an important constituency of government: its residential consumers.

We saw that the policy of promoting universal service in both Canada and the United States had a practical dimension. It was not simply a function of the marriage of industrial interests and political expediency. The actual implementation of monopoly regulation involving value-of-service pricing and system-wide cost averaging was an appropriate response to the presence of positive externalities associated with networks. In short, these policies could be justified on purely economic grounds. The rate-of-return formula for determining revenues may not have been optimal when measured against the economist's abstract standard of competitive markets, but it was an appropriate mechanism for balancing the interests of common carriers and consumers under monopoly conditions.

In both the United States and Canada regulatory bodies were created with responsibility for the communications sector. But the two models differed in a number of important respects. First and foremost, they differed with respect to the relationship between the agencies and political (elected) authority. We saw that the label of *independent* agency was much more applicable to the case of the FCC than that of the CRTC and its predecessors. Although rare, it was more common for a CRTC decision to be reversed by the federal cabinet than by the courts. Although direct government intervention of this type could not occur in the United States, the FCC was much more likely to be subject to reversals from the judicial branch of government. We see that the regulatory agency's relationship to the other branches of government has been determined by the larger structure of institutional arrangements associated with the founding of the nation-state and the political traditions and philosophies that inspired its creation.

Two-tiered regulation, where a company could be subject to two levels of regulatory authority, is a distinctly American phenomenon. Two-tiered regulation has meant a heightened role for the courts in resolving jurisdictional disputes between the states and the federal government in the United States. In Canada it is more appropriate to speak of divided or segmented jurisdiction, where a common carrier would be subject to either provincial or federal authority. Although more convenient for Canadian common carriers, this resulted in a fragmented and decentralized structure of regulatory authority that ultimately produced in a policy vacuum at the national level. How to create a national policy for the deregulation of the telecommunications sector when no agency had a priori national jurisdiction would become a major challenge to Canadian policy makers.

Under the U.S. Communications Act of 1934, the regulation of the broadcasting and telecommunications sectors was integrated in a single statute, and a single agency, the FCC, was created with responsibility for implementing the act. As a result of this institutional arrangement, telecommunications would have a high policy profile. In Canada broadcasting and telecommunications were always treated separately. Although creating policy for the broadcasting industry has been a major preoccupation of Canadian governments since the 1930s, the telecommunications sector was largely overlooked. For example, until 1976 the agency responsible for federally regulated common carriers was the CTC; moreover the relevant legislation was the Railway Act. For most of this century, therefore, telecommunications was a poor cousin to broadcasting in the realm of communications policy.

Although the institutional settings of the FCC and the CRTC were clearly different, they would share a common tension related to their dual roles as case-by-case tribunals and as de facto policy makers.

NOTES

1. The hypothetical case of monopoly price setting is taken from Richard Posner (1969, 551–552).

2. The Railway Act underwent numerous revisions. We refer here to the act as it stood in 1988. This version of the act has been reproduced by Peter S. Grant in a collection of laws pertaining to the communications secteur published in 1988 (Grant 1988).

3. We refer here to the Communications Act of 1934 as reproduced by Brenner in his *Law and Regulation of Common Carriers in the Communications Industry*. The page number refers the pagination found in Brenner's book; the article number refers to the Communications Act (Brenner 1992).

Conclusion to Part I

As the telephone industry matured in the early decades of the twentieth century, ownership of the industry became concentrated. Service at the level of both long distance and local service was provided by monopolistic carriers. In the United States, AT&T was the monopoly long distance carrier and dominated the local telephone market through its ownership of local operating companies throughout the country. Independent telephone companies entered the market after the expiration of the basic Bell patents in 1893 and 1894. At one point AT&T control of the local market in the United States fell to 51 percent. However, through an aggressive policy of acquisitions, the company would reverse this trend. By the 1930s, it controlled more than 80 percent of the local market through its ownership of local operating companies. In chapter two we described the Canadian industry structure based on dominant provincial carriers, a monopoly consortium for long distance (Trans Canada Telephone System [TCTS]), the predominance of Bell Canada (in terms of revenues and total subscribers), and a declining presence of independent carriers. The market for telegraph service was similarly concentrated, taking the form of an oligopoly in both countries.

How did monopoly come about? We know that the dominant carriers such as Bell Canada and AT&T employed a number of strategies to gain control of markets. Initially their dominant positions were due to their control of technology through patent rights. When these expired or were invalidated, competition from independent companies emerged in regions and markets not yet covered by the dominant firms. This occurred because of the high costs of setting up local networks. It was impossible for the dominant firms to secure financing that would enable them to develop everywhere and at once. But the dominant companies were not content to coexist with the independent companies. One of the most effective weapons in their arsenal was the power to refuse interconnection with their long distance network, or to charge exorbitant rates for such access. This had the effect of lowering the value of the telephone service provided by the independents. Also, the dominant companies were able to achieve control by acquiring their competitors.

As AT&T and Bell Canada increased their control of telephone markets, pressure grew to contain their emerging dominance of the industry. When the U.S. Justice Department threatened AT&T with an antitrust suit, the company decided to avoid litigation by moderating a number of its more questionable anti-competitive practices under the terms of the Kingsbury Commitment of 1913. Bell Canada reached a similar crossroads with the Ingersoll Telephone decision of 1911. In this decision the Board of Railway Commissioners ordered Bell to allow the interconnection of the Ingersoll network with its long distance network. Paradoxically, these interconnection agreements had a stifling effect on competition in the telephone industry. The terms for interconnection were generally favorable to Bell Canada and AT&T. Jean-Guy Rens notes that in exchange for interconnection Bell Canada gained the right to demand an additional fifteen cents per call, in excess of its normal rate, for calls originating from an independent company (1993a, 165). Under such favorable terms, Bell no longer had an incentive to compete with the independents in local markets. It would appear that this was the result sought by the commission. On the occasion of the decision, Chief Commissioner J. P. Mabee argued: "Although many think that competition is desirable, and it is in most cases. I have never found competition in telephony to be an interesting proposition" (165, our translation). A similar result was produced in the United States:

> [The Kingsbury Commitment] did indeed stop the growth of Bell's financial empire—for a few short years, in any event—*but it did nothing to promote competition in either telephony or telegraphy.* Local exchange monopolies were left intact, utterly free to continue to refuse interconnection to other local exchange companies. Bell's monopoly long distance service was reinforced: Bell would be required to interconnect with all local exchanges, but there was no provision for any competition—or interconnection—among long-distance carriers. Western Union was indeed spun off—but only to provide telegraphy, not telephony. (Kellog, Thorne, and Huber 1992, 16, emphasis added)

Under the complacent eye of regulators, AT&T and Bell Canada continued their strategy of acquisitions with devastating effects on the independents and the prospects for competition.

When we look back at this period it becomes clear that Bell Canada and AT&T could not have succeeded without the complicity of the government agencies responsible for regulating them. These agencies had the power to order the monopolistic, long distance carriers to allow interconnection with their networks on terms that would have been fairer to the local independent companies. For the most part, Bell Canada and AT&T were able to thwart the belated and lackluster efforts of the regulatory agencies to force interconnec-

tion. The result was market dominance and the creation of a de facto monopoly that lasted more than fifty years in Canada and the United States.

It is important to qualify the term "monopoly" as applied to the telephone
industry during this period. When we speak of monopoly in the Canadian and
U.S. telephone industries, we are not referring to a situation where a single
firm provided local service throughout the country, but rather, to a structure
of multiple providers each with a monopoly in a *geographic market*. In a given
geographic area only one company provided service. The effect of the interconnection agreements negotiated by the regulators was to balkanize the
telecommunications market. The following description of the U.S. market
was equally applicable to the Canadian industry structure of the period:

> The government solution, in short, was not the steamy, unsettling cohabitation that
> marks competition but rather a sort of competitive apartheid, characterized by seg-
> regation and quarantine. Markets were carefully carved up: one for the monopoly
> telegraph company; one for each of the established monopoly local telephone ex-
> changes; one for Bell's monopoly long-distance operations. Bell might not own
> everything, but some monopolist or other would dominate each discrete market.
> The Kingsbury Commitment could be viewed as a solution only by a government
> bookkeeper, who counted several separate monopolies as an advance over a single
> monopoly, even absent any trace of competition among them. (Kellog, Thorne, and
> Huber 1992, 16)

In chapter one we described this structure as the Canadian mosaic. In light
of our discussion here, it would probably be more accurate to describe it as
the Canadian *monopoly* mosaic.

The nature of the monopoly is probably best understood from the perspective
of the average consumer. First and foremost, the consumer did not have the option of choosing his or her local telephone provider. This was also the case for
long distance where Telecom Canada (TCTS) and AT&T were the sole providers
of this service in their respective countries. Moreover, all telephone sets were
provided to consumers by the local telephone company. Only equipment provided by the local telephone company could be connected to the network. From
the consumer's point of view there was no choice. There were no alternative
providers of telephone sets, local telephone service, or long distance telephone
service. In effect, from the consumer's point of view it was a service monopoly.

While Bell Canada and AT&T were building their monopolies, they were also
active in promoting the idea that their control of the industry was not a result of
anti-competitive behavior; rather, it was a natural result of the fact that the industry was a natural monopoly. The champion of these claims was Theodore Vail
under the slogan "One system, one policy, universal service."

A corollary to this was the view that the rise of Bell Canada and AT&T to
positions of dominance in their respective countries was a predictable outcome of natural monopoly. Similarly, the balkanization of telecommunications

markets into numerous geographic monopolies at the level of local service was also viewed as an outcome of natural monopoly. Our review suggests, however, that the type of industry structure that emerged had as much to do with the anti-competitive practices of the dominant carriers, coupled with the complicity of regulators, as it did with any inherent features of the technology that predisposed the industry to monopoly. To summarize, the historical record is inconclusive with respect to the question of natural monopoly. On the one hand, it is doubtful that the efficient operation of the network required that one company provide seamless service beginning with the provision of the basic telephone set and extending to the provision of the long distance telephony. On the other hand, the record is inconclusive as to whether the network for *local telephony* could have supported competition efficiently. Or alternatively, whether local telephone service could be provided most economically in conjunction with long distance. Moreover, these questions were largely unimportant during the period of natural monopoly. They would only become important in the 1970s when the liberalization and deregulation of telecommunications markets were being proposed.

Part II

FROM MONOPOLY TO COMPETITION:
THE DEREGULATION OF U.S.
AND CANADIAN TELECOMMUNICATIONS,
1946–1997

Introduction to Part II

An analyst considering the future of telecommunications in the 1950s and 1960s would have concluded that the network monopolies of the dominant carriers (AT&T and Bell Canada) were secure. The odds against competition prevailing over regulated monopoly would have appeared to be very slim. Let us consider for a moment the configuration of interests that were aligned against changes to the telecommunications sector.

Bell Canada provided service to approximately 60 percent of telephone subscribers and was the dominant force in the Trans Canada Telephone System (TCTS) long distance consortium (Department of Communications 1983a, 8). AT&T controlled 80 percent of the subscriber market and was the only provider of long distance telephone service in the United States. Bell Canada and AT&T were consistently among the top companies in their respective countries in terms of gross revenues. Not only did they dominate the telephone industry, they were powerful industrial actors providing a valuable infrastructure service to the entire economy.

AT&T was a key player in the U.S. military-industrial complex during the Cold War with the Soviet Union. Bell Laboratories, the research and development arm of AT&T, made a valuable contribution to national defense through the invention of technologies like the transistor. Western Electric, the manufacturing arm of AT&T, was a prime contractor in the development of the Nike Ajax guided aircraft missile. AT&T was also a key contributor to the U.S. Air Force's Distant Early Warning (DEW) air defense system based on radar technology. Northern Telecom, the research and development arm of Bell Canada, also made an important contribution to the defense of North America through its participation in the development of satellite technology and through the development of the DEW system along Canada's arctic border. Both AT&T and Bell Canada were important sources of scientific innovation and providers of technology to the military. As a result, they could usually count on the military to support them before Congress and Parliament whenever proposals were introduced to dismantle their vertically integrated businesses.

Perhaps most important, AT&T and Bell Canada were major employers. Prior to divestiture in 1984, AT&T was the country's largest private employer with one million employees (Simon 1985, 15). Bell Canada and its subsidiaries were major employers in Quebec and Ontario. In the case of AT&T, these employees were spread throughout the United States. They could be found in every county, city, and town where a Bell System local operating company operated a telephone exchange. These were not just employees, they were members of local communities. They were citizens with the right to vote. Most important, they were members of very powerful and well-organized unions. Employees and their unions understood only too well that competition would result in cuts in employment and diminished job security. These unions could be counted on to work with management to oppose any challenge to AT&T's status as monopolistic provider.

Finally, it is important to remember that the telephone network that emerged under regulated monopoly was highly accessible, affordable, and reliable. Telephone users were generally satisfied with the service they received. Because the system was not broken, there was no consumer movement to fix it. There may have been some dissatisfaction with the heavy-handed, bureaucratic culture of the telephone companies, but the problem was not serious enough to warrant appeals for a major restructuring of the industry on the part of consumer groups.

Given these factors, it was extremely unlikely that major structural change would be introduced to the industry. And yet, against formidable odds, change did occur. It is precisely because the forces opposing change in the telephone industry were so powerful and so entrenched, that the transition from monopoly to competition makes this an appropriate and worthwhile object for study.

Chapter four begins our analysis of the deregulatory movement by examining the seeds of change and how the ideas of analysts and theorists of regulated industries (primarily economists) prepared the way for a radical reorganization of the telecommunications sector. Chapters five and six analyze the deregulation of U.S. telecommunications beginning with the liberalization of the terminal equipment market and culminating in the liberalization of the local exchange. Chapters seven and eight consider the same events as they occurred in Canada.

4

The Power of Ideas: The Beginning of the End of Monopoly in Telecommunications

4.1 INTRODUCTION

It is generally acknowledged that the Federal Communications Commission's (FCC) 1968 *Carterfone* decision marked the first decisive breach of AT&T's monopoly in the United States. Specifically, the decision opened the way for the attachment of non–AT&T equipment to the Bell System. In Canada the first decisive step towards competition occurred almost ten years later in 1976 when the Department of Communications introduced its Terminal Attachment Program (TAP). The program set standards for the attachment of certified equipment by unregulated equipment providers to the facilities of federally regulated telephone companies. These events may have marked the unofficial commencement of liberalization, but the groundwork for change was being prepared at least a decade earlier by researchers (primarily economists) who challenged the fairness, the efficiency, and the overall benefits to society of economic regulation of natural monopolies. In this chapter we will recount the history of these ideas and this research. We will see that ideas such as "competition," "deregulation," "liberalization," "regulatory capture," "structural reform," and "contestable markets," can be potent weapons when endorsed by powerful corporate interests as part of a campaign for radical change.

4.2 THE PROBLEM WITH REGULATION: SOME PROMINENT FAILURES

By the mid-1970s the inefficiencies of a number of regulatory agencies (most of them created during the New Deal period) began to draw public attention in the United States. The performance of the railroad, trucking, airline, and energy industries, all subject to price and entry regulation, was being questioned in the United States. Although the telephone industry was not a particular focus of criticism, the FCC was also subject to heightened scrutiny due to its status as a regulatory agency operating in the price and entry mode. In an article entitled "The Dead Hand of Regulation," James Q. Wilson (1971) described some of the perverse effects of Interstate Commerce Commission (ICC) regulation of the transportation sector in the United States. He noted that the price to ship something from California to Maine by railroad was the same as that from Idaho to Pennsylvania. This was the case even though the routes could differ by as much as one thousand miles. More specifically, he states:

> These and other anomalies result from the fact that the ICC has felt it necessary to set rates on an item-by-item, point-to-point basis, partly because it sought originally to protect the farmer (at the expense of the manufacturer) and partly because the railroads themselves practiced this form of price discrimination in the days before they had much competition. The ICC decided to strike a compromise between the interests of farmers and those of the railroaders by tinkering with existing prices rather than by starting all over. What began as tinkering wound up as a set of 75,000 separate rate schedules and an administrative procedure for handling (in 1962) over 173,000 proposed tariffs submitted by truck and barge as well as rail firms. (44–45)

As a result of the case-by-case, administrative approach to regulation, the ICC was spending its time making fine-grained distinctions "between horses for slaughter and horses for draught, between sand used to make glass and sand used to make cement, and between lime used in industry and lime used in agriculture" (Wilson 1971, 45). But the problem was not simply one of excessive rule making and administrative overkill.

Railroads competed with trucking, another industry subject to regulation by the ICC. In its attempt to regulate competition across industry barriers, the ICC introduced further distortions to the market. As could be expected, the trucking industry responded by competing with the railroads on long-haul routes where rates had been set artificially high as a result of ICC regulation. Although one would expect that the trucking industry would be most efficient on short-haul routes, regulation meant that the industry was drawn to long haul where profits were greater. Clearly, this was not an efficient use of resources. Wilson notes that the ICC responded, not by freeing the railroads so they could compete, but by introducing more regulation that prohibited the trucking industry from undercutting railroad prices on those routes (Wilson 1971, 45). This case illustrates one of the principal failings of regulation: Its response to regulatory distortions tends to be the introduction of more regulation. As regulation engenders more regulation, it becomes relevant to question whether the benefits of regulation continue to outweigh the costs associated with the imposition of more rules and regulations.

Problems were not limited to the transportation sector. The U.S. Federal Power Commission's (FPC) decision to favor consumers with low rates for natural gas eventually produced a gas shortage. This resulted in the rationing of gas in several states in the early 1970s. Gas was so cheap that there were few incentives for the natural gas companies to explore for new sources of gas. Moreover, at least one economist, Paul MacAvoy, had predicted as early as 1962 that FPC pricing policy would lead to these shortages (Donahue 1971, 197).

Finally, the Civil Aeronautics Board (CAB), responsible for regulating the airline industry, was the focus of particular public scrutiny and criticism. To cite one example, it cost less per mile to fly from Los Angeles to San Francisco than from New York to Washington. Carriers on the California route charged an average $26 per mile compared to an average $42 per mile from New York to Washington. The price differential was explained by the fact that as an intrastate route, Los Angeles to San Francisco was not subject to federal regulation by the CAB (Breyer 1979, 588). Stephen Breyer, in an article entitled "Analyzing Regulatory Failure: Mismatches, Less Restrictive Alternatives, and Reform," notes:

> Classical price regulation also tended to protect the firms already in the industry and their market shares. The CAB allowed no new firms into the industry after 1950. The market shares of the major airlines in 1975, compared with those in 1938 when the industry was 1/435 its later size, showed remarkable stability. Finally, there appeared to be little incentive to perform more efficiently. Airlines that had lower costs were not allowed to seek more business by charging lower prices or, for that matter, by offering, wider (and fewer) seats. (588–589)

It would be unfair and inaccurate to argue that the regulation of the transport, energy, and airline industries was in a state of crisis. Nor would it be accurate to use the term "regulatory failure" to describe what was occurring in these industries. Clearly, regulation had its costs and inefficiencies, and increasingly the shortcomings of regulation were drawing the attention of politicians and the press. However, it was one thing for regulation to surface as a public issue, it was quite another for dissatisfaction with regulation to reach a level sufficient to induce radical reform. Writing in 1971, Wilson expressed the general scepticism of the period that important changes to regulation were in the offing:

> Neither industry nor the agencies have much need to fear major reforms, or at least reforms that reduce the item-by-item discretionary regulation that now exists. Quite the contrary. The reform impulse, except among economists who specialize in the problems of better use of regulated resources, is now of an entirely different sort—increased regulation, increased discretion, more numerous challenges to existing corporate practices. . . . With respect to rate-fixing agencies, a move toward greater reliance on market forces in industries where competition is adequate is also unlikely. Not only would the agencies oppose it, but some groups will lose in the short run from any deregulated rate: Consumers might have to pay more for natural gas, oil companies might earn less from crude oil, and some shippers would pay more in rail and truck rates. Everyone who stands to pay more will naturally oppose deregulation, and those who will pay less (consumers, future generations, some firms) are typically not organized to seek such benefits—if, indeed, they are even aware of them. (57)

Before the politicians could take up the challenge of change, a strong and detailed case for deregulation had to be made by the proponents of competition. Essentially, there were two analytical approaches to the study of regulated industries: a traditional approach, inspired by administrative law and practiced largely by lawyers; and an emerging economic approach inspired by neo-classical economic theory and proffered by economists. The traditional approach, which preceded the arrival of economists and the economic analysis of regulated industries, emphasized the importance of fairness and due process in regulatory procedures; these studies generally concluded with appeals for the *reform* of regulation. The economic approach was generally more critical of regulated industries. As could be expected, the economic analysis of regulated industries resulted in studies demonstrating that competition would be more efficient than regulation; essentially, it was argued that the costs of regulation outweighed its benefits. These studies generated appeals for *structural changes* that would deregulate the industries and introduce competition.

4.3 WHAT TO DO ABOUT THE REGULATORY AGENCIES: THE REFORM APPROACH

It was not long after the creation of the first independent regulatory commissions in the United States that governments began to review their performance and consider proposals for their reform. In 1937 the President's Committee on Administrative Management described the independent agencies as "a headless fourth branch of the Government, a haphazard deposit of irresponsible agencies and uncoordinated powers" (Anderson 1983, 435). Other attempts at reform included the following: the Attorney General's Committee on Administrative Procedure (1941); the Hoover Commission (1949 and 1955); James Landis's *Report on Regulatory Agencies to the President Elect* (1960); and, finally, the report of the President's Advisory Council on Executive Organization (1971)—also known as the *Ash Report* (435).

The Attorney General's Committee on Administrative Procedure (1941) led to the Administrative Procedures Act of 1946. As the title suggests, the act established standard rules and procedures to ensure that all parties before a regulatory commission would receive fair treatment. The rules concerned such aspects of legal *due process* as the right to ample notice (so parties could prepare their arguments), procedures to ensure a fair hearing, and the opportunity for review of decisions (Wilson 1971, 41). Wilson contends that the act did not herald a new era in administrative law. Instead, the principal effect was to make "regulation slower and more costly rather than fairer" (41).

The *Ash Report* submitted to the White House in 1971 was typical of the administrative reform approach. The report noted that the independent regulatory agencies lacked flexibility and were not adapting to changing industry conditions. Decision making was too slow. There was no coordination of policy between agencies. The *Ash Report* proposed five changes in administrative structure to deal with the problem:

1. Limit the number of commissioners as much as possible—where possible replace them with a single administrator
2. Make the agencies more responsive by making the heads of agencies members of the President's administration
3. Increase the efficiency of the process of deciding individual cases by creating an Administrative Court
4. Rationalize the distribution of functions among agencies for better inter-agency coordination (Noll 1971, 9)

Consistent with the administrative reform approach, the *Ash Report* did not question the existence of the regulatory agencies. They performed a necessary function. However, they were in need of reform if they were to keep pace with the requirements of a changing economy.

In *The End of Liberalism*, Theodore Lowi (1979) explored the repercussions for the political life of the nation of delegating so much authority for public decisions to agencies operating in the administrative mode. In a comprehensive, influential, and highly critical study, Lowi argued that the administrative approach to public policy was based on a public philosophy that he referred to as "interest-group pluralism." Essentially, interest-group pluralism was based on the premise that the group (as opposed to the individual citizen with a right to vote) was the fundamental unit of political activity. Traditional liberal political philosophy held that the free competition of ideas articulated by individual citizens would result in the best ideas rising to the top where they would eventually gain public support. The political process was essentially self-correcting, resembling the supposedly "self-corrective" mechanisms of the marketplace. Interest-group pluralism put a new, but reassuring, spin on the old liberal philosophy: The rise of powerful interest groups, with full-time lobbyists in Washington, did not represent a fundamental threat to democracy. The self-correcting process still worked, but at a different level: the level of the interest group.

Lowi was highly critical of the new public philosophy, arguing that there was no evidence that interest-group politics were self-corrective and would serve the public interest. He contended that the new public philosophy was based on a naive and idealized concept of the group. Finally, he argued that the new public philosophy failed to take into account that competition between interest groups for public support would not be fair. Given the resources and economic power of certain actors, it would be difficult for the process to meet the criteria of perfect competition.

In his book, Lowi argued that the trend toward delegating authority to the independent regulatory agency was symptomatic of the crisis in public philosophy.

Clearly, it was the role of the independent regulatory agency to arbitrate conflicts between the competing interest groups. This placed the regulatory agency
at the center of the political life of the nation. Lowi was particularly critical of
the case-by-case approach to administrative rule making that we described in
the previous chapter. He stated:

> Most of the administrative rhetoric in recent years espouses the interest-group lib-
> eral ideal of administration by favoring the norm of flexibility and the ideal that
> every decision can be bargained. Pluralism applied to administration usually takes
> the practical form of an attempt to deal with each case "on its merits." But the ideal
> of case-by-case adjudication is in most instances a myth. Few persons affected by a
> decision have an opportunity to be heard. And each agency, regulatory or nonregu-
> latory, disposes of the largest proportion of its cases without any procedure at all,
> least of all a formal adjudicatory process. In practice, agencies end up with the
> worst of case-by-case adjudication and of rule-making. They try to work without
> rules in order to live with the loose legislative mandate. (1979, 299–300)

The solution, according to Lowi, was to tighten and make more precise the legislative mandates of the agencies, and to encourage the agencies to make early
and frequent rules. This would neutralize the power of interest groups who
tended to benefit from the bargaining process that occurred as a result of the
case-by-case approach (1979, 300).

It was not until the late 1970s that regulatory reform began to preoccupy researchers and policy makers in Canada. As late as 1978 Richard Schultz observed, "Compared to their American counterparts, Canadian independent regulatory agencies have been almost completely ignored by both academic
students of public administration and government inquiries into administrative
issues and problems" (1978, 291). This began to change in the late 1970s when
a number of studies examining the regulation of the transport and energy sectors were launched.[1] The low level of interest in Canada probably reflected the
fact that the "independent" regulatory agency model was less applicable to
Canada than it was to the United States. In fact, only three federal agencies
could be said to approximate the "independent" agency model: the Canadian
Transport Commission (CTC), the National Energy Board (NEB), and the
Canadian Radio-television and Telecommunications Commission (CRTC). At
least one study concluded that the low level of public interest in the regulatory
process could be explained by the historical tendency in Canada to favor state
intervention when necessary (Doern et al. 1975, 212). Other factors included
the traditional Canadian deference to authority and a political culture that often
relied upon "elite accommodation"[2] to resolve issues of public import (212). As
we noted in chapter three, one must be careful not to attribute the same degree
of independence to the Canadian agencies as their American counterparts.
However, in the Canadian case, the fact that the agencies were less independent did not make them immune from controversy or criticism. Indeed, the un-

easy, and at times contradictory, relationship between the regulatory commissioners and government would make the commissions an ideal object for reform.

Writing in 1979, Hudson Janisch reviewed the status of Canada's independent regulatory agencies. His analysis confirmed that the central issue facing these agencies was their relationship with government and government policy. The issue was neatly summarized in the report of the Royal Commission on Corporate Concentration, which stated the following:

> For example, is there any acceptable way of reconciling the principle of regulatory independence with political control? Should government be able to overrule, direct or otherwise interfere with the decisions of independent regulatory boards exercising powers given to them by Parliament and, if so, on what grounds? To what extent and in what manner should executive interference be subject to the supervision of Parliament? What should be the proper scope of interest group participation in the regulatory rule-making process, and how can this be assured and controlled? How often, in what way and in what respects should regulatory boards be accountable to the legislatures that created them? Is Parliament able to examine critically the policies and decisions of regulatory boards . . . (Royal Commission 1978, 404)

The issue in Canada was not so much the problem of regulatory failure, but rather, the relationship between the agencies and elected authority. Critics charged that this relationship was plagued with inconsistencies and contradictions. Although problems of due process, flexibility and adaptability, and overprotection of the regulated companies were noted, there was no ground swell of public concern with the overall performance of the agencies. The model worked. It simply was in need of reform so that it could work better.

In chapter three, section seven, we described the relationship between government and the regulatory agencies in Canada. We noted that the parliamentary model allowed for a greater degree of control of the regulatory commissions by elected authority (cabinet). Although this was an accurate theoretical description of the relationship, in actual practice, the mode of oversight and the frequency of government intervention in the activities of the regulatory agencies varied considerably. To illustrate the problem, Janisch (1979) has referred to a matrix developed by Schultz (1977) based on two forms of cabinet review (active and passive) and two powers of action (positive and negative) (table 4.1).

Table 4.1   Modalities of Government Intervention and Oversight of Regulatory Bodies

	Active	*Passive*
Positive		
Negative		

The active/passive axis takes into account whether cabinet (technically the Governor in Council) has the authority to intervene upon its own initiative or whether it must simply respond to a petition from one of the parties. When the cabinet has the authority to intervene upon its own initiative, without appeal from petitioners, it is said to have *active* powers of intervention. In cases where the cabinet's authority is reactive, it has *passive* powers of intervention. The positive/negative axis refers to the types of action that the cabinet can take. In cases where legislation allows cabinet to reverse a decision, and substitute a new one, it is said to have *positive* powers. When cabinet is limited to simply returning a decision to the agency for reconsideration, its powers are *negative*.

The problem with the Canadian regulatory system was the apparently arbitrary manner in which particular legislation authorized cabinet to intervene in the different regulatory jurisdictions. For example, the Broadcasting Act (1967–1968) conferred "active-negative" powers on cabinet with respect to the CRTC decision to license a broadcaster, but no power of review in cases where the CRTC simply refused a license (Janisch 1979, 63). When making decisions in the telecommunications sector, the same agency, the CRTC, was subject to "active-positive" control (63). Licensing decisions of the CTC were subject to "active-positive" control, but the minister of transport had "passive-positive" powers of review with respect to the licensing of airlines (63).

Janisch summarized the inconsistencies and contradictions of the Canadian system in the following manner:

> No underlying rationale can be found for this allocation of review power. Why, for instance, should the cabinet have the power to completely reverse a telephone rate determination of the CRTC, but have no power at all over a gas rate set by the NEB? [National Energy Board] Why has the government only a limited power to review a decision of the CRTC to grant a licence, but apparently unlimited power to "vary or rescind" any decision of the CTC? [Canadian Transport Commission] Why should the [M]inister of [T]ransport, who could be in a position to ensure that air licensing decisions are made consistent with government policy, have his appeal power dependent on the initiative of the parties involved? (1979, 63)

The apparently arbitrary distribution of review powers resulted in a confused situation that threatened to undermine the legitimacy of both the regulatory agencies and cabinet. The problem was particularly acute in the area of policy formation discussed in chapter three. There was a pressing need to get away from the case-by-case approach so that the agencies could concentrate on long-term policy formation. However, in order to be successful, a new framework was required that would clarify the roles that cabinet and the agencies would assume in the policy-making process. New legislation was required that would: (1) prescribe the conditions under which cabinet or a department of government could issue broad policy directives to an agency; and (2) delegate responsibility for policy formation to the agency, subject to predetermined and limited powers of review.

The reform approach in both the United States and Canada was based on the dual premise that economic regulation was both necessary and appropriate. It was necessary to protect the public interest from the abuses of market power. It was appropriate as an instrument for the control of monopoly because it could be counted on to deliver most of the benefits of the market, albeit through other means. Although necessary and appropriate, it was, nevertheless, perfectible. If economic regulation could be made fairer, more predictable, and more accountable to the public, it would be able to realize its objectives. In the United States this could be accomplished by making the legislative mandates of the agencies more precise and by encouraging the agencies to move away from the case-by-case approach in order to rely more on general rules that would guide the behavior of industry. Organizational changes that would limit the number of commissioners and make the heads of agencies more accountable to the executive branch would make the agencies more accountable to government. These and other reforms were required to ensure that powerful interests (representing either the companies they regulated or consumers of their products) did not capture the regulatory agencies. In Canada there was less dissatisfaction with the performance of the agencies themselves, and less concern with potential for capture by powerful interests. Paradoxically, it was the parliamentary system of government that prescribed a greater degree of elected, political control of the agencies, which created uncertainty and, even, the potential for abuse.

4.4 ENGINEERING CONSENT FOR DEREGULATION: BUSINESS USERS AND THE ECONOMIC APPROACH

In 1985 the Brookings Institution published the first comprehensive study of the deregulation of the U.S. airline, transport, and telecommunications industries. Entitled, *The Politics of Deregulation* (Derthick and Quirk 1985), the study sought to identify the forces that contributed to U.S. deregulation, to evaluate the different theories of how and why change had occurred, and to develop its own theory of how change occurs in large, complex institutions. The authors of the study, Martha Derthick and Paul J. Quirk, credited the community of pro-competitive economists with laying the foundations of the deregulatory movement, and through their efforts, of having played a major role in the eventual success of the movement. They noted, "Economists had begun making the bullets of pro-competitive regulatory reform fifteen years before politicians found them to be usable in particular battles they wished to fight" (56). Elsewhere they stated: "We are convinced that except for the development of this academic critique of policy, the reforms we are trying to explain would never have occurred . . ." (36).

The economic crisis of the 1970s set the stage for a new alliance between academic economists and the U.S. business community. As we shall see in this

section, neoclassical economists had already begun their attack on regulated industries in the early 1960s. In the 1970s their work bore fruit. Corporate interests under the institutional umbrella of groups such as the Business Roundtable
and the American Enterprise Institute (AEI) used the work of the free market
economists to bolster a new social policy agenda that was antagonistic to state
planning and intervention in the economy. In his book *Policy-Planning Organizations: Elite Agendas and America's Rightward Turn* (1987), Joseph G. Peschek
demonstrates that these groups were heavily funded by corporate interests and
had a comprehensive program for policy change. He notes that there was a:

> massive inflow of corporate funds into AEI coffers during the 1970s, which enabled
> the launching of new study projects and publications, and a huge outreach program
> involving the media, universities, public conferences, and congressional testimony
> and staff contact. Much of this agenda-building and opinion-molding was directed
> at domestic economic policy. The breadth and perspective of AEI work is evident in
> the areas of democratic capitalist ideology, regulatory studies, macroeconomic
> analysis and recommendations, as well as in the prominent figures active in the
> AEI. (191–192)

Dan Schiller in his *Telematics and Government* (1982) has chronicled the link between the growing demands of corporate users and regulatory change. He states:

> Business users had to press themselves upon regulators simply because the regu-
> lated telecommunications industry itself was not meeting their dynamic telematic
> needs. Both the terms and the costs of use for developing equipment and services
> often set the traditional telecommunications industry and the largest business users
> at odds. As a result of this spiraling conflict, the entire shape of the traditional
> telecommunications industry has been transformed and, in our era, business users
> of telematics have become a decisive policymaking force. (3)

Agenda building and opinion molding are by nature slow and laborious
processes. This is particularly the case when their objective is nothing less than
a redefining of corporate-state relations. Extensive funding over a considerable
period of time is a prerequisite. But change also requires ideas. In the case of
the telecommunications sector, funding was provided by suppliers of telecommunications equipment seeking new markets, potential competitors to the common carriers, and high-volume corporate users of telecommunications (Schiller
1982). The ideas were provided by the neoclassical economists of the Chicago
School of economics, and were circulated almost a decade before being taken
up by corporate interests as part of a concerted effort to force a change in policy that would reduce the costs of corporate telecommunications.

When economists set their sights on the institution of economic regulation, they fixed their attention on the *effects of regulation*; this contrasted with
the administrative reform approach, which emphasized the process and pro-

cedures of regulation. One of the first attempts to quantitatively measure the effects of economic regulation was undertaken by George Stigler and Claire Friedland. In 1962 they published the results of a study that examined the impact of regulation on rates for electricity in the United States. They compared the price of electricity in states with regulatory boards with prices in states with no regulation. Their analysis of various years from 1912 to 1937 led them to conclude that regulation did not result in lower prices for consumers. The fact that regulation did not produce lower rates for consumers placed into question an important pillar of the public interest theory of economic regulation: the assumption that regulation protected consumers from monopoly prices (Stigler and Friedland 1962).

In chapter three, section five, we described the rate-of-return formula employed to determine the revenue requirements of regulated firms: R = E + r × B. Recall that E (expenses), comprised the costs of labor and the cost of depreciated capital (equipment). In 1962 Harvey Averch and Leland L. Johnson developed an economic model to describe the likely behavior of a regulated firm, operating within the rate-of-return framework. When the model was applied to the behavior of a hypothetical firm it produced the following result: "If the rate-of-return allowed by the regulatory agency is greater than the cost of capital but is less than the rate-of-return that would be enjoyed by the firm were it free to maximize profit without regulatory constraint, then the firm will substitute capital for the other factors of production and operate at an output where cost is not minimized" (Averch and Johnson 1962, 1053). Simply stated, this meant that the rate-of-return formula encouraged the firm to over-capitalize, that is, to use too much capital vis-à-vis labor and other factors of production. According to the authors, the rate-of-return formula created incentives for public utilities to adopt a cost of capital that was below the real social cost of capital. This produced an uneconomic investment in modernization and plant expansion when labor or other factors of production could have produced an equivalent result at lower cost. Moreover, the problem, as framed by the study, was not so much dishonest or corrupt management, but a revenue formula that encouraged the misallocation of resources. It was to be expected that the public utilities would take advantage of a bias in the revenue formula that favored investment in capital equipment as a means toward increasing their net return.

Harold Demsetz, in an article entitled "Why Regulate Utilities?" (1968), questioned one of the fundamentals tenets of the theory of natural monopoly: economies of scale in production lead to monopoly, and, ultimately, to monopoly pricing. According to the theory of natural monopoly, a natural monopoly exists when one producer can supply the market more efficiently than several producers. This occurs when production exhibits economies of scale. Economies of scale are "factors which cause the average cost of producing a commodity to fall as output of the commodity rises" (Bannock, Baxter, and Davis 1989, 129). For example, when economies of scale are present, a company could double its production without doubling its costs. This situation is typical of production in industries where there is a heavy initial cost of capital, for example, for physical plant and equipment. Once this initial investment has been made, the cost of additional inputs necessary for production are low. This results in an average cost that declines as production increases. To compete under these conditions would require a major capital investment, similar to that of the incumbent firm. But the benefits to society of declining average costs, which accompany increased production, would not be realized to the same extent because production would be distributed between two firms. It is generally acknowledged that these conditions prevail in public utilities where the requirement to provide a direct connection with the consumer entails a high initial capital cost.

In his article, Demsetz did not question whether public utilities and other natural monopolies benefited from economies of scale. Nor did he challenge the supposition that competition in the presence of significant economies of scale would be uneconomic. Rather, he argued that there was no basis for the conclusion that economic concentration (monopoly), resulting from economies of scale, would result in so-called monopoly prices. This assertion was based on a somewhat specious argument that there was no basis in *economic theory* for making the leap from economic concentration to *monopoly prices* (i.e., prices set artificially high due to the absence of competition). Demsetz summarized his argument in the following way: "The important point that needs stressing is that *we have no theory that allows us to deduce from the observable degree of concentration in a particular market whether or not price and output are competitive*" (1968, 61, emphasis added).

We can say in retrospect that Demsetz's article was more of a footnote in the analysis of regulated industries than a major chapter. However, it is representative of the kind of critique that emerged in the late 1960s when pro-competitive economists began to set their sights on economic regulation. The article signals the beginning of what can be called the *basic* or *fundamental* critique of public utility regulation. By this we mean that the goal of this type of analysis was to question the fundamental suppositions of economic regulation, as opposed to exploring how regulation could be made more efficient or fairer. In this instance, Demsetz is challenging the rationale for price regulation: If there is no basis for believing that the monopolist will charge monopoly prices, then price regulation is unnecessary. In chapter three, section five, we demonstrated that there are incentives for the monopolist to price discriminate and charge monopoly prices as a way of increasing profits. It matters little that Demsetz does not accept or is unaware of these arguments. As an example of the emerging pro-competitive critique of economic regulation, the article signaled the new direction that research into public utilities was taking—a direction that would challenge the usefulness, indeed the very necessity, of monopoly regulation.

In a very lengthy article published in the *Stanford Law Review*, Richard Posner (1969) elaborated what is probably the first comprehensive cost-benefit analysis of natural monopoly regulation. Entitled "Natural Monopoly and its Regulation," the article identified all the shortcomings, perverse effects, and uncalculated costs of regulation in order to conclude that the disadvantages of regulation outweigh its benefits. Although the article is too lengthy to review in detail, Posner's critique of *profit regulation* will be reviewed as an example of his approach.

One of the cornerstones of natural monopoly regulation is the control of company profits, usually through the rate-of-return formula described in chapter three, section five. In building his case against regulation, Posner reviewed some of the pitfalls of profit regulation. He noted that during periods of declining operational costs, the regulated firm may exceed its prescribed rate of profit. Because regulation is usually slow to react to changing conditions, it may take years before a new, lower rate-of-return is imposed on the monopolist to curtail these profits. Also, it is possible for the regulated firm to conceal monopoly profits through "adroit accounting" practices (1969, 594). The rate-of-return approach requires that decisions be made concerning which assets are to be included in the calculation of the rate base. These decisions are largely a matter of judgement. In assigning costs it is particularly difficult to determine which costs should be assigned to which services. For example, how much of the cost of the telephone set should be assigned to local service and how much to long distance (595). The way that costs are assigned has an impact on the determination of revenues. Posner also questioned whether profit regulation would encourage the regulated firm to engage in wasteful expenditures such as excessive salaries and benefits to managers. He stated:

> Regulation may encourage other wasteful expenditures. Management can react in two ways to a ceiling on profits. It can charge the price that will return the allowed profit and no more. Or it can charge the monopoly price but convert the forbidden profit into increased cost. The latter is the course that managerial self-interest could be expected to dictate. Since executive salary is formally a cost, the firm would not appear to exceed its profit constraint. However, the transfer of profits to management in the form of exorbitant salaries is one kind of evasive maneuver that a regulatory agency is likely to be able to detect. Consequently, conversion of monopoly profits into management perquisites would be likely to assume less transparent forms. (601)

Finally, Posner questioned the efficacy of profit regulation by suggesting that one could not limit a firm's ability to increase its profits without affecting the "monopolist's incentive to efficient, progressive operation" (598). The challenge for regulators must be to control profits, but at a level that will not destroy the company's incentives for innovation and efficient operation. Posner was skeptical that regulators would be able to identify the optimum level.

Posner's article goes on to question virtually all aspects of natural monopoly regulation. The critique even includes a challenge to the principle of system-wide price averaging that was the basis for universal service. For Posner, price averaging was nothing more than an uneconomic subsidy flowing from one group of users to another. From Posner's perspective, it would be more useful to give disadvantaged individuals a subsidy of money (e.g., in the form of a tax deduction) rather than a subsidy of service. Moreover, according to Posner, internal subsidies do not reach their target group effectively since "all too often, the principal beneficiaries turn out to be members of the affluent middle class" (1969, 608).

Posner concluded his analysis of the effects of regulatory control in the following manner: "there are different degrees of justification for the various regulatory controls, but in no case do the benefits clearly outweigh the costs. There is no pervasive case for the regulation of specific rates in, or entry into, natural monopoly markets; yet these have been important areas of regulatory activity, whose principal result has been to promote inefficient pricing and to create unjustified barriers to entry and competition" (1969, 618).

Probably the most frequently expressed criticism of economic regulation was the charge that the regulatory commissions were subject to *capture* by powerful interests. The capture theory can be broken down into two schools of thought: the theory of capture by regulated firms (also referred to as clientism), and the general interest group theory of capture. (Hereafter, we will refer to the theory of capture by regulated industries as the *client* theory of capture.) The client theory of capture contends that the regulatory commissions are subject to control by the industries they regulate. According to this theory, regulation is introduced—usually against the protests of industry—to protect the public interest. Eventually, however, the regulators begin to identify with the companies they regulate, who become their clients. In other words, they come to confuse the *public interest* with the interests of the companies they regulate. The general interest group theory of capture charges that regulatory commissions are subject to capture by powerful interests, but does not limit the field of powerful interests to the companies subject to regulation. According to the general theory of capture, the regulatory commissions may be "captured" by any number of groups. For example, by consumer groups, or a subset of consumer groups such as business or residential consumers.

The client theory of capture predates the arrival of the economic critique of public utility regulation. Before the economists began to examine the effects of regulation, political scientists had already begun to question whether the regulatory commissions were serving the public interest. For example, Marver H. Bernstein, in *Regulating Business by Independent Commission* (1955), argued that the independent commissions "have lacked an affirmative concept of the public interest; they have failed to meet the test of political responsibility in a democratic society; and *they tend to define the interests of the regulated groups as the public interest*" (296, emphasis added). Although Bernstein did not use the term "capture" to describe this unsatisfactory state of affairs, his words reflect the suspicion (one that has plagued public utility regulation since its inception) that in the course of the evolution of economic regulation the regulators will come to identify with the industries they regulate.

A number of reasons have been offered to explain why regulatory tribunals would be susceptible to capture by their client industries. Public utility regulation requires that the regulated company provide detailed information on its operations to the regulator. The regulator is completely dependent on the regulated company for this information, since without it there would be no accurate basis for setting prices. By controlling the flow of this information and by "structuring" it in an expedient fashion, the regulated company may be able to control the regulator. It is argued, that given this informational dependency, it is only a matter of time before the regulator will begin to see the world through the perspective of the companies it regulates. Other factors explaining the pro-industry bias include the following: the lack of attention by politicians to the activities of the commissions, the use of appointments to the regulatory boards to gain favor with affected industries, and the ability of the regulated industries to promise job creation in exchange for favorable treatment by regulators and politicians (Derthick and Quirk 1985, 92).

In 1971, Stigler married economics to the general interest group theory of capture (1971). This produced the first general theory of regulated industries based on an economic approach. Stigler argued that regulation could be treated as a commodity subject to the laws of *supply* and *demand*. Governments had the power to supply regulation. Industry, and other likely beneficiaries of regulation, created a demand for it. To predict when and where regulation would be introduced, it was simply necessary to understand the interaction of supply and demand in the market for regulation. Because the potential benefits of regulation to different groups will vary, it is logical to assume that the demand for regulation will be distributed unevenly in society. It is not necessarily the group with the highest visibility or greatest numbers that will prevail in appeals to government to introduce regulation. Rather, it is the group with the greatest self-interest.

In theory, a small group, which stands to reap a large economic benefit from regulation, may prevail over a numerically larger (and potentially more politically powerful) group, which will suffer a small economic loss as a result of regulation. In the case of regulation, supply meets demand where the relative benefits of regulation are the greatest. If the economic benefits of regulation to producers are significant and the economic costs to consumers are marginal, the economic theory of regulation suggests that supply will meet the demand of the producers. It should be noted that Stigler's theory is just that, a theory. As is the case with any theory, its validity resides in its power to explain and predict events. It is important to note that the theory was not derived from an empirical study of the regulated industries. As a general theory of economic regulation, it was an attempt to explain the underlying dynamics applicable to all cases of economic regulation.

Although overly ambitious, untested, and largely unverifiable, the theory reached a wide audience and had a significant impact on thinking about regulated industries. The success of the theory probably had more to do with the approach, which had the aura of scientific method, together with the general ascendancy of economic analysis of the period, than it did with the theory's actual ability to explain or predict events. Essentially, Stigler was restating what political scientists had already postulated in the general interest group theory of capture, but Stigler's theory came with the promise that it could be verified through rigorous analysis using the proven methods of the discipline of economics. Most important, here was a theory of regulation, articulated by an economist for other economists, that could be employed to explain the *failures* of regulation.

Stigler's theory of economic regulation marked a new chapter in the analysis of regulated industries for two reasons. First, the theory signaled a fundamental shift in the debate on regulated industries: from an approach based on the reform of regulation to an approach that questioned who really benefited from regulation. Moreover, it did so in a manner that seriously undermined the prevailing public interest theory of regulation. If the supply-demand theory of regulation was correct, there was little basis for holding that economic regulation was serving the public interest. As we will see, such thinking was the basis for an unlikely alliance between neo-classical economists and consumer groups who came together in the mid-1970s to challenge the regulatory establishment on the grounds that regulation was putting the interests of monopolistic corporations over those of consumers. Second, it signaled the arrival of economics as the dominant paradigm in the analysis of regulated industries. No longer would economists limit their analysis to price comparisons of regulated vs. unregulated industries, or abstract speculation on the relationship between economic concentration and monopoly pricing. Stigler's work cleared the way for economists to expand their analysis into fundamental questions concerning the origins (genesis) and functionality (social/political function) of monopoly regulation. These were questions that had previously been the privileged domain of historians and political scientists.

We saw that Demsetz (1968) based his critique of monopoly regulation on the argument that there were no grounds (in economic theory) for assuming that the monopolist would charge monopoly prices. Stated in other terms, Demsetz was contesting the prevailing wisdom concerning the likely *behavior* of an unregulated monopolist. Demsetz was not the only economist to challenge monopoly regulation on the basis of an alternative theory of the likely behavior of firms. In 1982 William Baumol, John Panzar, and Robert Willig proffered a radical new theory of industry behavior with significant implications for the traditional theory of monopoly. In *Contestable Markets and the Theory of Industry Structure*, they argued that competition is not a function of the number of firms operating in a market. Rather, competitive results can be realized even when only two (or even one) firms are producing for a given market. This was the case so long as the monopolist was operating in what they termed a "contestable market."

According to Baumol, Panzar, and Willig, a market is contestable if *potential* entrants face no inherent disadvantages vis-à-vis incumbent firms. In order to meet this requirement, the following conditions must be met: competitors must have access to the same technology as the incumbents and at the same price; there must be no legal restrictions on entry; the incumbent must not benefit from any government subsidies; there must be no special costs to be born by the competitor that are not shared by the incumbent; and most important, entry must be reversible—there must be no sunk costs (xx). (Sunk costs are costs, typically capital investment in equipment, that a firm would be unable to sell in the event that it withdraws from the market.) Under "contestable" conditions, where there is a "potential" for competitive entry, the incumbent monopolist can be expected to behave like a firm operating in a competitive market. In the words of the authors:

> Using contestability theory, economists no longer need to assume that efficient outcomes occur only when there are large numbers of actively producing firms, each of whom bases its decisions on the belief that it is so small as not to affect price. What drives contestability theory is the possibility of costlessly reversible entry. . . . In contrast [to the purely competitive model] both incumbents and potential entrants in a contestable market recognize their power over prices and realize that they cannot sell more than consumers demand at given prices, without bidding market prices down. Consequently, a contestable market need not be populated by a great many firms; indeed, contestable markets may contain only a single monopoly enterprise or they may be comprised of duopolistic or oligopolistic firms. (xix–xx)

The case for contestable markets is built on highly technical arguments that are only of interest to economists. For the purposes of our review, it is sufficient to note that Baumol, Panzar, and Willig proffered their theory as one that could be "extraordinarily helpful in the design of public policy" (1982, xxii). Clearly, the theory of contestable markets has important implications for public utility regulation and the theory of natural monopoly, since it postulates that monopoly does not necessarily result in so-called monopoly behavior. Stated in other terms, the theory contends that monopoly in a contestable market is fully compatible with the type of behavior that one could expect from a firm operating in a competitive environment. An unregulated monopolist operating in a contestable market could be expected to operate at the same level of efficiency, provide service at equivalent prices, and have the same incentive to innovate, as a producer operating in a multifirm market.

The theory of contestable markets is controversial, even among economists. The requirement that there be no sunk costs (that entry is reversible without penalty) is particularly demanding, since it is difficult to imagine a situation where all investment related to startup would be reversible. Needless to say, once the theory had been advanced the next step was to develop a set of analytical tools that could be used to test whether a given market met the conditions of contestability set out in the theory. Again, this generated debates and challenges to the theory that are primarily of interest to economists. For the purpose of our analysis, it matters little whether the theory of contestable markets is viable. It is sufficient to note that it was an important theory, credible to many economists, that generated considerable research and analysis. Most important, the theory influenced the debate on how to deregulate public utilities by fortifying the position of the pro-competitive economists.

The problem of natural monopoly can be broken down into three concerns: (1) what is natural monopoly—that is, when is a monopoly natural and how do we identify natural monopolies?; (2) what is wrong with natural monopoly—that is, why does it need to be regulated?; and (3) what type of regulation is most appropriate? Economists were less interested in exploring the first question than they were in dissecting the second and third. This was a judicious approach, since the first question is without import so long as one is unable to present a case (theoretical or empirical) for the need to regulate, together with arguments to demonstrate that regulation will be effective in resolving the problem. If one is not able to demonstrate that natural monopoly is in some way detrimental to the good of society, then it matters little how one goes about identifying it.

As we have seen, the economists' first line of attack was to reduce the second question (what is wrong with monopoly) to the issue of monopoly pricing. On the basis of Stigler and Friedland's comparative study of pricing, two conclusions may be drawn: either unregulated monopolists do not necessarily charge monopoly prices; or (and less plausibly) the regulated and unregulated public utilities were both charging monopoly prices and the commissions were not doing their job to prevent it. In the first instance the economic raison d'être for monopoly regulation is called into question. The second conclusion encourages speculation on the behavior of the regulator, as opposed to that of the monopolist. The Averch and Johnson study casts doubt on the effectiveness of the mode of regulation: question three. It suggests that rate-of-return, which has generally been conceived as the most appropriate method of regulation, produces unanticipated, perverse effects. Ultimately, it raises the question: Is there an efficient, effective way to regulate public utilities? Demsetz's reflections on the relationship between monopoly and monopoly pricing are the theoretical counterpart to the work of Stigler and Friedland. Posner's work juxtaposes the second and third questions in an effort to bring them together as part of a single equation. His analysis cuts two ways: on the one hand, it minimizes the onerous consequences of unregulated monopoly; on the other, it magnifies the costs and shortcomings of monopoly regulation. In the new equation, a revised assessment of the perils of monopoly is weighed against a skeptical view of the benefits of regulation; this is done in order to cast doubt on the merits of regulation. Posner's assessment of the costs and benefits of monopoly regulation challenges an unexamined premise of economic regulation, that regulation is not burdened by significant costs or other intrinsic liabilities that, on balance, would compromise its effectiveness.

Stigler's supply-demand theory of regulation is atypical in that it has little direct bearing on the three questions that form the basis of the debate on monopoly regulation. Essentially, Stigler's theory is a theoretical exploration of the role that competing *interests* play in determining the structure of regulation. In this respect it is akin to the political analysis of regulation as a social institution. However, Stigler's work does have implications for the three questions. If he is correct, then there are legitimate grounds for questioning whether the public interest theory of regulation is little more than a rhetorical sham, a smokescreen designed to deflect attention away from the play of economic interests that are the real force behind economic regulation.

Finally, the theory of contestable markets was an attempt to generalize and expand the scope of neo-classical market analysis by redefining a fundamental tenet of neo-classical economics: the theory of perfectly competitive markets. According to the theory of contestable markets, the laws of market economics can be extended to include industries that are "potentially" competitive. The field of economic analysis could then be expanded to include industries that were previously viewed as subject to market failure. The theory of contestable markets is designed to eliminate, or at the very least obscure, the line that separates regulated from unregulated industries. As such, it calls into question the first question: What is a natural monopoly? The theory of contestable markets responds to this question by contending that monopolies are not that different from competitive markets. This contention is the case once we understand the effect that "potential" competition will have on the behavior of firms.

Our review of the economic critique of natural monopoly regulation is a small, but representative, sample of the research that laid the theoretical foundations for U.S. deregulation. Hundreds of books and thousands of articles have been published, and continue to be published, on the subject of competition and regulated industries. We have elected to review the pioneering literature that had a seminal influence on U.S. deregulation. It is important to note that in addition to being economists, the authors we have reviewed were members of the Chicago School of economics. The Chicago School is known for its staunch neo-liberal approach that favors free markets above any form of government intervention at the microeconomic level. The rigid preference for markets over government regulation is reflected in the work of Milton Friedman, who, besides being a Nobel laureate, is generally considered the leader of the Chicago School.[3] Friedman expounded what would become the Chicago School approach to natural monopoly as early as 1962 in his book *Capitalism and Freedom*. He argued that when there are "technical conditions [that] make a monopoly the natural outcome of competitive market forces, there are only three alternatives that seem available: private monopoly, or public monopoly, or public regulation. . . . I reluctantly conclude that, if tolerable, private monopoly may be the least of the evils" (28). Monopoly was tolerable, because eventually competition from the fringes of the market could be expected to challenge the monopolist. On the other hand, once regulation was entrenched, it might be extremely difficult to displace. As a free market economist, Friedman had more faith in the power of markets to eventually reverse monopoly, than he did in the ability of regulation to exit and allow competition in response to changing industry conditions. If one took the long view, unregulated monopoly was the lesser of two evils. Essentially, this would be the position of the Chicago School economists. Over the long term, and given the shortcomings and perverse effects of regulation, it is preferable to suffer monopoly than submit to regulation. Of course, as we have argued in chapter three, section three, such an approach is profoundly ahistoric as it ignores centuries of social policy designed to curb the excesses of the market and ensure fairness in economic relations.

4.5 FROM IDEAS TO PUBLIC POLICY: DEREGULATION GAINS THE SUPPORT OF POLITICIANS

Notwithstanding the regulatory failures described in section two, the reform of public utilities was not a high priority among voters and politicians in the early 1970s. The critique of economic regulation proffered by the Chicago School economists would probably have fallen on deaf ears had it not been for the economic crisis of the 1970s and the efforts of large corporations to use the crisis as a pretext to promote deregulation. But this was only half of the equation. Although the corporate agenda had been set, there still remained the problem of engineering consent both within government and with the general public. As we shall see in this section, support for deregulation would come from an apparently unlikely source: the consumer movement.

Uncommonly high levels of inflation characterized the 1970s. Throughout the 1950s price inflation averaged only 2.2 percent annually, and it rose slightly to 2.6 percent during the 1960s. However, during the 1970s the yearly rate more than doubled to 7.5 percent; some years it reached double-digit levels (Peschek 1987, 49). One of the principal causes of inflation was the 1973 decision of the Organization of Petroleum Exporting Countries (OPEC) to stop consulting with the American and European oil companies before raising the price of their oil exports.

Energy costs, however, were not the only causes of inflation. A number of analysts had begun to speculate on the relationship between chronic inflation and increased social welfare expenditures by government. In the United States, expenditures by government in social welfare programs comprised 32.7 percent of government spending in 1955. By 1965 the percentage had risen to 42.2 percent; in 1975, 57.9 percent. As a percentage of the gross national product (GNP), social welfare expenditures in the same three years were 8.6 percent, 11.7 percent, and 19.9 percent (Peschek 1987, 194). The cost of government during a period of chronic inflation was beginning to surface as a public issue. Although the performance of regulatory institutions was not initially a public concern, the costs of government regulation, of all kinds, would be. High inflation combined with the rising cost of government were two of the defining political issues of the 1970s. Before we can understand how public debate on inflation and the high cost of government (and the relationship between the two) shaped the pro-competitive deregulatory movement, a third component must be grasped: the rise of consumerism.

Ralph Nader is generally acknowledged to be the founder of the U.S. consumer movement in the 1970s. Trained as a lawyer, Nader led the crusade for safer automobiles in the mid-1960s with the publication of his book *Unsafe at Any Speed: The Designed-in Dangers of the American Automobile* (1965). In 1971 he founded an organization called Public Citizen Incorporated. As a crusader for consumer rights, Nader fought for stricter health and safety laws in the food industry, including more stringent controls on meat and poultry production and the use of pesticides. The consumer movement was generally skeptical of the relationship between big government and big business, contending that government agencies were serving corporate interests and not the public interest.

The inflationary spiral of the 1970s presented a special challenge to the regulatory commissions. Higher costs of capital (interest rates) and higher labor costs compelled the public utilities to raise their prices. Under rate regulation these rate increases went to the various state regulatory commissions for review. Persistent inflation resulted in repeated applications by the public utilities for rate increases—applications that were opposed by consumer groups. The regulatory commissions were characteristically slow in reviewing these requests: it frequently took several years to reach a decision. As a result of regulatory lag, the public utility companies lost money (Horwitz 1989, 204). When the commissions finally sanctioned rate increases, the suspicions of consumer groups that the regulators were serving the interests of big business were confirmed.

Regulatory reform and consumerism probably came together for the first time in the Administrative Practice and Procedure Subcommittee of the Judiciary Committee, chaired by Senator Edward Kennedy of Massachusetts. Stephen Breyer,[4] who was special council to the subcommittee, urged Kennedy to undertake hearings on the reform of the airline industry. Derthick and Quirk have speculated on the motives behind Kennedy's initial interest in the following terms: "Kennedy was attracted to the subject because he saw in it a response to consumerism, the political movement that addressed the interests of the consumer against big business or, in the case of regulatory regimes, the combination of big business and big government" (1985, 41). The hearings were a major success, receiving wide coverage in the U.S. press. As result of the hearings, the competence of the Civil Aeronautics Board (CAB; the agency responsible for regulating the airlines) was called into question. For example, the CAB was unable to justify the differentials in fares for trips of equivalent distance described in section two. Incredibly, the CAB was unable to present a convincing case in support of rate and entry regulation (44).

As Kennedy expected, consumer groups testified in opposition to airline regulation. Nader, in testimony before the subcommittee was particularly critical of the CAB. He stated, "people are repulsed by arrogant and unresponsive bureaucracies serving no useful public purpose, and they are looking to this Congress to get on with the national housecleaning job that is needed. Can you think of a better place to start than the Civil Aeronautics Board?" (quoted in Derthick and Quirk 1985, 51). Derthick and Quirk argue that as a result of the airline hearings, the emerging critique of economic regulation was made accessible to the general public. Not surprisingly, the airline industry would later become one of the first industries to undergo deregulation. The Airline Deregulation Act of 1978 put an end to the economic regulation of the U.S. airline industry and, ultimately, to the CAB.

One of the defining characteristics of the pro-competitive deregulatory movement was the way it drew support from both the liberal left and the conservative right. This was even true at the level of the executive branch of government. In 1974 Gerald Ford succeeded Richard Nixon, who resigned at the beginning of his second term as president. Ford, a Republican, was defeated by Jimmy Carter, a Democrat, in the presidential elections of 1977. In 1981, after serving a single term as president, Carter was defeated by the Republican candidate Ronald Reagan. In spite of their different party affiliations, Ford, Carter, and Reagan all supported deregulation.

Ford, a Republican, embraced deregulation from a traditional conservative position. When he took office, inflation was the leading domestic issue. His administration quickly came to the conclusion that big government was a major cause of inflation; reducing big government was the remedy. In order to identify areas of government that were ripe for reform, Ford established a presidential task force, the Domestic Council Review Group on Regulatory Reform. The task force met once a week bringing together advocates of reform from the different branches of government (Derthick and Quirk 1985, 46). During the Ford administration two major reform bills were submitted, dealing with the airline and trucking industries.

Although regulatory reform did not figure prominently in his campaign for president, Carter embraced the movement following his election. Whereas Ford supported a broad program of reform and deregulation that would significantly reduce government's role in the economy, Carter's approach was more restricted. His approach resembled Kennedy's consumer-driven reform. It was largely on the basis of consumer arguments that Carter threw his support behind bills that were already moving through Congress to deregulate the U.S. airline and trucking industries. Carter softened industry resistance to these bills by appointing pro-competitive commissioners to the regulatory agencies. Slowly, in a piecemeal fashion, the regulatory commissions began to introduce competition to these industries. This incremental approach to deregulation discouraged industry opposition to legislative reform. As *administrative* deregulation advanced, the trucking and airline industries realized that it was politically untenable to propose legislation that would reverse pro-competitive reforms. Moreover, they began to favor the idea of legislative action that would put an end to the uncertainty resulting from incremental deregulation.

Reagan's approach to the reform of big government resembled that of Ford, his Republican predecessor. By the time he was elected in 1981, Reagan had been campaigning for years against big government, overregulation, and bureaucratic excess (Derthick and Quirk 1985, 30). Whereas Carter's support for deregulation was shaped by the ideology of consumerism, Reagan's was driven by the ideology of free enterprise capitalism. Reagan's program of reform went beyond the existing conservative/liberal consensus on economic regulation that had resulted in the deregulation of the trucking and airline industries. He set his sights on the health, labor, safety, and environment "bureaucracies" that had been created in the 1960s and 1970s. According to the Reagan view, not only were these bureaucracies responsible for increasing social welfare expenditures, they were also throttling the creative, innovative spirit of American business.

The groundwork for Reagan's attack on big government had been prepared years before by conservative policy organizations such as the American Enterprise Institute (AEI), the Heritage Foundation, and the Institute for Contemporary Studies. These organizations are financed primarily by grants from private corporations. Often referred to as "think tanks," their principal activity is to fund studies conducted by academics and other analysts into political issues that are tailored to meet the interests of their conservative clientele. Ultimately, their goal is to influence government policy. The crusade against big government was not just about out-of-control spending and burgeoning bureaucracy. More fundamentally, it entailed an ideological project to rehabilitate capitalism. At one level this was done by promoting the association between free markets and democracy, that is to say, between personal freedom and the market place. Peschek (1987), in his study of policy-planning organizations, has summarized the role played by the AEI in these efforts in the following terms:

> the AEI has established a Project on Democratic Capitalism to promote a positive view of the economic system. It is significant that this project not only rests on arguments about the material advantages of a market economy, but also emphasizes the moral-cultural component of democratic capitalism. Long the target of criticism from adversary intellectuals, the AEI views the culture of capitalism as the seat of freedom of ideas, tolerance, social concern, and the search

for talent in all classes. . . . In short, what the AEI has helped to develop is a
more attractive argument for capitalism, which is depicted not as a Hobbesian
jungle or a barren plain of mechanistic rationality, but as a precondition for
achievement of the finest liberal ideas. (224)

Reagan's election marked a turning point in the movement for pro-competitive
deregulatory reform. What began as a critique of economic regulation in the
1960s, had taken on the attributes of a political crusade, orchestrated by some of
America's largest corporations, with strong ideological overtones. By the late
1980s support for smaller and less intrusive government had gathered a wide constituency in the United States, which in many cases transcended party affiliation.
Consensus between conservatives and liberals was probably strongest with respect to the need to pare government bureaucracy: the rule-bound system of
forms, reports, and other controls that were throttling American industry. How to
achieve this was another matter, since the bureaucracy in question was itself the
product of recent congressional action.

The reader will recall the three eras of regulation discussed in chapter
three: Progressive, New Deal, and Great Society. To a considerable degree, it
was the Great Society program of regulation—which was introduced to deal
with the social consequences of business in the areas of health, safety, and
the environment—that was responsible for the growth of government bureaucracy. Since the 1980s, the conservative, Republican approach to the
problem of big government has been to emphasize a reversal/removal of Great
Society regulation. The liberal, Democratic approach has been more prudent,
emphasizing the need for administrative reform while maintaining protections for consumers and citizens.

How is this discussion relevant to an understanding of the pro-competitive
deregulation of public utilities? We see that the program of reform that was initiated during Ford's administration as a solution to the problem of chronic inflation was transformed during the Reagan years (and in the years to follow) into a
political challenge to the Great Society regulation of the period 1965–1977. For
the free enterprise ideologues, the primary thrust of deregulation, its real objective, must be the elimination of Great Society regulation created to deal with
business externalities. But this has proven to be difficult, because many of the
protections in the areas of health, safety, and the environment have proven to be
popular with voters. Instead, deregulation has been most successful in reversing
regulation that was introduced during the New Deal period (1930–1938). In
other words, notwithstanding the loud and vociferous voice of the political right
on the need to unshackle American business and reduce burgeoning social welfare expenditures, deregulation has been most successful as a campaign to reverse fifty-year-old price and entry regulation *that was not the immediate or principal cause of the inflationary spiral of the 1970s.* Moreover, while American
business has been the driving force behind the deregulatory movement since the

1970s, deregulation has been most successful in sectors of the economy where the principal industrial actors, and their unions, actually opposed change: airlines, trucking, and telephone. The irony of this situation has not been lost on Robert Britt Horwitz, who has reclaimed it as the title of his book on telecommunications, *The Irony of Regulatory Reform: The Deregulation of American Telecommunications* (1989). Horwitz has ably summarized this curious state of affairs in the following terms:

> Nonetheless, after all the rhetoric and all the efforts to roll back social regulation, those agencies have not been dismantled. Their statutes remain intact, and legislative campaigns to weaken specific mandates (such as the Clean Air Act) have failed. An effort to deregulate the social agencies internally through a conscious pattern of slow-down, policy reversal, and nonenforcement has had some, but as the record of judicial review shows, also limited success. A central irony of the deregulation phenomenon is that only those industries whose principal players generally desired continued regulatory oversight were deregulated. . . . Furthermore, the dissolution of economic regulation took place largely *before* Reagan came to power. (265)

4.6 CONCLUSION

The reversal of decades of monopoly regulation that began in the United States in the 1970s was a uniquely American phenomenon. It was the result of an historically unique set of circumstances, particular to the United States, with few parallels elsewhere in the world—including Canada. It is important to remember that the general critique of public utility regulation, and the specific critique of telecommunications regulation, were by-products, specific effects, of a larger anti-government, deregulatory movement in the United States—a movement that was financed and coordinated by a diverse but powerful constellation of corporate interests. At one level, the deregulatory movement was a political response to the inflationary pressures of the 1970s. But it was also an ideological response, orchestrated by the conservative right, to the Great Society social programs of the period. Although the Great Society programs were the principal target of the conservative deregulatory movement, the movement was most successful in challenging public utility regulation, the product of an earlier period of regulation. This has been one of the ironies of regulatory reform. As a political movement, the attack on public utility regulation was unique in that it garnered support from both the left and the right of the political spectrum. This unusual coming together of conservative, pro-competitive interests, and liberal, pro-consumer groups was key to the success of the deregulatory movement.

We saw that in Canada during the same period the focus of debate and discussion was on how to reform regulation. Generally, there was less concern about regulatory boards being subjected to capture by their client industries, than their being subjected to undue political influence by cabinet. Reform was required to clarify and to harmonize the relationship between cabinet and the various regulatory boards. Most important, there were calls for changes to legislation that would clarify the respective roles of cabinet and the regulatory boards in setting of long-term policy. Eventually, the deregulatory impulse would begin to drive policy making in Canada, but this would only occur after deregulation had become firmly entrenched in the United States.

Deregulation was the product of a clash of interests (producers vs. consumers, monopolists vs. their competitors, regulators vs. private corporations and the U.S. Congress), but it was also a product of the clash of ideas. The economic critique of public utilities set the stage for a shift in policy favoring competition. But the ascendancy of economics did not occur in a void. These ideas were cultivated and promoted by important economic interests. The proponents of pro-competitive deregulation were able to win the day over the partisans of administrative reform because they had the backing of powerful, privately funded, policy organizations such as the AEI and the Heritage Foundation. These ideas received a positive response by both Congress and the executive branch because they were supported by both consumer and business groups. As public opinion began to shift in favor of deregulation, regulators at the FCC, and the other regulatory boards, gradually began to adjust their policies to favor competition. How this occurred in the United States will be chronicled in the next chapter. In chapter seven we will analyze the Canadian response to American deregulation.

In at least one respect the success of the deregulatory movement is not surprising. The regulation of natural monopoly by the state is a potent symbol of the failure of markets, and by extension, of the failure of the field of economics. It is difficult to reconcile the discipline of economics and its doctrinaire commitment to competition and the power of markets, with the presence of natural monopoly and the necessity of economic regulation. It was possible for economists to overlook, even tolerate, economic regulation during the economic collapse of the Great Depression. As economic stability returned in the years following the Second World War, it appeared that the problem of under consumption and its corollary of over production, the impetus for the Great Depression, had been solved. This created opportunities for economists to reapply their theories and models to regulated industries such as telecommunications. We have seen that the economic critique of public utility regulation was comprehensive. By challenging the rationale for monopoly regulation, by questioning the efficacy of regulatory measures, by obscuring the boundaries that separated monopoly from competitive industries, and by casting doubt on the interests that were being served by regulation, little room was left for halfway measures of reform. By undermining regulation and promoting the idea that competition in regulated industries was sustainable, the economists were also able to bolster the standing of their profession, the field of economics.

In this chapter we have suggested that the economic critique of public utilities was a precondition for the pro-competitive deregulation that followed. We have presented the economists' position verbatim. This was not done out of a conviction that the neo-liberal economic critique should be considered the final word on network economics. It is not. In our opinion, the neo-liberal economic approach is not a sufficient basis for building a public policy for telecommunications. A policy for telecommunications must take into account: (1) the existence of network externalities; (2) the problem of how to create incentives for investment in network facilities (which may be put at risk as a result of the uncertainty created by competition); and finally, and most important, (3) the social welfare requirements of large segments of the population who as a result of competition may drop out of networks. These issues will be addressed in later chapters where we will analyze the limitations of the neo-liberal economic approach to network policy.

NOTES

1. Janisch notes that the Law Reform Commission of Canada sponsored a number of agency studies as part of its Administrative Law Series. These included: Bruce G. Doern, *The Atomic Energy Control Board* (Ottawa: Minister of Supply and Services 1976); Alastair R. Lucas and Trevor Bell, *The National Energy Board* (Ottawa: Minister of Supply and Services, 1977); and Hudson Janisch, *The Regulatory Process of the Canadian Transport Commission* (Ottawa: Minister of Supply and Services 1978). Also, beginning in 1977 the Institute for Research on Public Policy announced that it would undertake a number of studies on government regulation (Janisch 1979, 46).

2. The theory of elite accommodation contrasted with the American theory of interest-group pluralism. It was an attempt to explain Canadian political behavior in terms of the interaction of a closed network of organizational leaders, an elite (Doern et al. 1975, 192).

3. Friedman won the Alfred Nobel Memorial Prize for economics in 1976.

4. In 1994 Breyer, nominated by President Bill Clinton, was appointed to the U.S. Supreme Court.

5

Step-by-Step toward Deregulation
in the United States

5.1 INTRODUCTION

The story of U.S. deregulation is essentially the story of attempts to break AT&T's vertically integrated monopoly. Between 1913 and 1982 there were three government-initiated attempts to break AT&T's monopoly using antitrust law. All of them were resolved through negotiated settlements that preempted a ruling by the courts. We will consider each of them in this chapter. On a second front, AT&T's monopoly was undermined by a series of decisions issued by the Federal Communications Commission (FCC) and the courts relating to its control over key sectors of the telephone market: terminal equipment and long distance. As we will see, the courts and the FCC were often at cross-purposes with respect to AT&T's right to monopolize these sectors of the telecommunications market. Eventually, both sectors were liberalized. We will also consider the case of satellite communications, a sector that from its inception emerged under liberalized market conditions with the support of both the FCC and the executive branch of the U.S. government.

What will emerge is a portrait of AT&T as a stubborn and vigorous defender of its monopoly, a company that used all the legal-procedural and competitive tools at its disposal to quash entry. Beginning in the 1960s, the FCC became a cautious supporter of competition, but only at the margins of AT&T's business. We will see that this approach was not sustainable. It was impossible to introduce a little "stimulative competition" at the periphery of AT&T's business because such a policy could only be based on a totally arbitrary separation of competitive from non-competitive sectors of the market. Each time the FCC drew the line on where monopoly should stop and competition should begin, it found itself reversed by the courts, technology, or other factors. Finally, we will see that the wildcard in the game of pro-competitive deregulation was the courts. On numerous occasions the courts intervened either to open markets, or in the wake of the AT&T divestiture, to attempt to manage the introduction of competition in a manner unanticipated by the FCC and the telephone companies.

5.2 THE CONSENT DECREE OF 1956: AT&T DEFLECTS A SECOND ANTITRUST CHALLENGE TO ITS MONOPOLY

In chapter one, section three, we described how AT&T successfully resolved the first antitrust challenge to its monopoly by entering into a negotiated settlement known as the Kingsbury Commitment (1913). This enabled the company to avert a lengthy court battle and a decision by the courts. A second, more serious challenge to AT&T's monopoly had its origins in the mid-1930s after authority for regulating the telephone industry had passed to the FCC under the Communications Act of 1934. Shortly after its creation, the FCC undertook an investigation into the industry in order to determine if more legislation was required. The investigation took several years, cost $2 million, and mobilized three hundred researchers (Brock 1981, 179). One of the principal recommendations of the initial report concerned the relationship between the Bell System operating companies and their equipment supplier, Western Electric, which was owned by AT&T. At the time, Western Electric was the sole supplier of telephone supplies to the operating companies. The report recommended that the operating companies introduce competitive bidding for telephone supplies. In effect, the report challenged AT&T's vertically integrated structure. In the final report, which proposed only minor changes to the 1934 Communications Act, this recommendation was quashed. But it resurfaced ten years later as the basis for a Department of Justice (DOJ) antitrust suit against AT&T and Western Electric, launched on January 14, 1949. Not coincidentally, the lead attorney in the 1949 antitrust case had been a researcher with the FCC in the 1930s.

The 1949 antitrust suit did not directly challenge AT&T's network monopoly, which encompassed local and long distance telephone service. Rather, it sought to dissolve AT&T's relationship with its manufacturing arm, Western Electric. The suit charged AT&T and Western Electric with unlawful restraint of trade resulting in a monopolization of the telephone equipment market. AT&T and Western Electric were able to control the equipment market largely through their practice of acquiring and licensing patents. The suit charged, "the basis for the success of AT&T in occupying and controlling almost the entire telephone operating and manufacturing fields and for its extensive activities outside the telephone industry has been due largely to its development and exploitation of patents" (*United States v. Western Electric Co.*, No.17-49 (D. New Jersey, Jan. 14, 1949)). In order to remedy the situation, the DOJ proposed that AT&T relinquish its ownership of Western Electric. It was proposed that Western Electric's market power be reduced by dismantling the company into three separate companies. Finally, if successful, the suit would have put an end to all restrictive agreements between AT&T's long distance operations, its local operating companies, and Western Electric (Brock 1981, 187).

According to Gerald W. Brock, AT&T responded to the suit by preparing a three-part plan. First, it began to prepare its legal defense. Second, the company solicited the support of the U.S. Department of Defense to build its case that the integration of research and development with manufacturing was important to the defense of the nation. Third, AT&T began to explore the possibility of a negotiated settlement that would enable it to maintain its monopoly. In spite of the support that AT&T received from the Defense Department, the company was unsuccessful in having the case dismissed. However, in consideration of AT&T's contribution to the Korean War, the case proceeded at a slower pace during the war years. Finally, in January 1956 AT&T's negotiated approach prevailed. The settlement, which became known as the Consent Decree of 1956, was entered into by AT&T and the U.S. attorney general (*United States v. Western Electric Co.*, Trade Cas. (CCH) ¶68,246 (D.N.J 1956)). The centerpiece of the agreement was a more liberal approach to the sharing of patents with AT&T's manufacturing competitors. Access to existing and future Bell System patents would be granted to all comers on a non-exclusive basis. AT&T would provide licenses with technical information on how to manufacture the equipment for a reasonable fee (Kellog, Thorne, and Huber 1992, 203). Thereafter, Western Electric would restrict its manufacturing to the production of equipment used in the provision of telephone service. Finally, AT&T agreed to limit its activities to the *business of common carriage*. What did AT&T gain in return for these concessions? Under the terms of the settlement, AT&T was able to maintain its ownership of Western Electric. The vertical integration of research and development, equipment manufacture, and network operations (both local and long distance) was maintained.

The 1956 Consent Decree was ironically labeled the "Final Judgement." Although this optimistic view may have reflected the consensus of the principal parties to the negotiations, the settlement was anything but final. During the twenty-eight years it was in force, the FCC had to deal with numerous challenges to AT&T's monopoly. Many of these challenges were the product of commercial opportunities created by new technologies, such as microwave transmission, satellites, and computer-based data transmission. When AT&T was slow to develop these technologies, it created opportunities for the development of private, corporate networks that were outside the public (common carrier) system. But these networks were never completely isolated from the public network. Although private, they often required some partial transmission using network components provided by AT&T. Moreover, it was a short step from private networks owned and operated by the corporations that used them, to closed networks, offered by third parties on a commercial basis to these corporations. It became increasingly difficult for the FCC to arbitrate disputes between AT&T and the alternative providers who sought to operate on the margins of basic, public telephone service. Finally, it is important to emphasize that the Consent Decree of 1956 did not immunize AT&T from further antitrust prosecution under the Sherman Antitrust Act (1890). Although this may seem surprising, it is understandable when one takes into account the principles underscoring decades of antitrust jurisprudence. These will be discussed in the final section on the AT&T divestiture.

The Consent Decree of 1956 remained in effect, for twenty-eight years, until 1982 when AT&T settled another antitrust suit with the DOJ (*see* section nine). During this time, it was the legal basis for AT&T's vertically integrated business. But it was more than this. By setting limits on the types of businesses AT&T could engage in, it provided the basic blueprint for the telecommunications industry structure. The Consent Decree sanctioned, or at the very least gave the appearance of sanctioning, AT&T's dominant position in the telecommunications industry, since the company was not compelled to dismantle its vertically integrated business. What company could possibly aspire to compete with AT&T on these terms? But it did not resolve, once and for all, the problem of AT&T's monopoly. That is to say, it did not settle the problem of AT&T's right to compete unfairly in markets for future telecommunications services. Nor did it completely discourage potential competitors.

In the next sections we will consider how competition was introduced to markets for *terminal equipment* and *long distance* services. We will see that AT&T's right to monopolize these markets was not etched in stone. Rather it was subject to the discretion of the FCC and the courts. At times the FCC sided with AT&T in rulings that were designed to protect its monopoly. On other occasions, the FCC encouraged competition, but usually on the margins of AT&T's business. The courts also played a role. In general the courts were less tolerant of AT&T's anti-competitive maneuvers than the FCC. On a number of occasions the FCC found that the courts reversed its decisions supporting monopoly. This had an impact on subsequent FCC decisions, obliging the commission to be more cautious in its support for AT&T domination of the industry.

5.3 *HUSH-A-PHONE* AND *CARTERFONE:* AT&T LOSES ITS MONOPOLY IN THE PROVISION OF TERMINAL EQUIPMENT

The term "terminal equipment" designates a device (or devices) used to terminate a communication channel and whose purpose is to adapt the channel for use by a user, whether a machine or a person (Simon 1985, 132). In the 1950s the most common examples of terminal equipment were the standard telephone set, switchboards, and teletypewriters. Today, terminal equipment includes such devices as fax machines, modems, answering machines, pagers, and cellular phones.

AT&T's strategy to maintain its monopoly was based on an end-to-end (telephone set to telephone set) control of the telephone system that was intolerant of any form of competition. As part of this strategy, the company refused to allow the attachment of terminal equipment from other suppliers to its network. The company was so zealous in the protection of its monopoly that it opposed all forms of attachment to the network, including any modification to the terminal equipment it provided to users. Since 1928 the Hush-A-Phone Corporation had been selling a plastic, snap-on device (with no electrical components) that a telephone user could place over the mouthpiece of the telephone handset in order to reduce background noise found in an office environment. In 1949 AT&T complained to the FCC that the Hush-A-Phone device had the potential to lower the quality of its network by muffling a person's voice. Seven years later the commission finally responded to the complaint, ruling that foreign attachments presented a potential threat to the integrity of Bell's network; thereafter, all foreign attachments to terminal equipment would be reviewed on a case-by-case basis by the commission. Moreover, the commission determined that the Hush-A-Phone device itself would "be deleterious to the telephone system and injure the service rendered by it" (FCC 1955, at 420).

Not content with the commission's ruling, the Hush-A-Phone Corporation appealed the case to the federal court, where the FCC decision was reversed. The court found no support for the commission's (and AT&T's) contention that the Hush-A-Phone device would adversely effect the telephone network. But the court went further to establish the right of telephone subscribers to use the telephone in "ways which are privately beneficial without being publicly detrimental" (*Hush-A-Phone Corp. v. United States*, 238 F.2d 266 (D.C. Cir. 1956)). In retrospect it is clear that AT&T overreacted to Hush-A-Phone. The device clearly did not pose a threat to AT&T's network. AT&T's real goal in opposing Hush-A-Phone was to discourage any and all future competitors to its terminal equipment market, and more importantly, to garner the FCC's support for such a policy. Although AT&T's behavior was understandable, the FCC's decision to align itself so closely with AT&T in opposition to such an innocuous device is more difficult to comprehend. The court's ruling in favor of Hush-A-Phone was an embarrassment to the FCC, and a warning to the commission that if it continued to shelter AT&T from all current and potential competition it would be exceeding its authority and risking the specter of renewed antitrust.

The *Hush-A-Phone* decision did not open the floodgates to competition in the terminal equipment market. But it did establish an important principle concerning the subscriber's right to use the telephone, and other terminal equipment, in ways not prescribed by the company so long as such use was not harmful to the network. It would take over ten years for the implications of this principle to be felt in the market for terminal equipment. This occurred in 1968 in a landmark ruling known as the *Carterfone* case (*In the Matter of the Use of the Carterfone Device in Message Toll Telephone Service*. 13 F.C.C., 2d 240, 13 R.R. 2d (1968)).

In 1959 the Carter Electronics Corporation of Texas began manufacturing a two-way radio device that could be connected to the telephone network and used for mobile telephony. The Carterfone enabled a caller (using the telephone land-wire network) to be linked with someone at a remote (mobile) location. The device consisted of a connecting device (a cradle) that the telephone headset was placed in. The cradle was linked to a radio transmitter that broadcast its signals to a portable radio receiver. Essentially, the device extended the range of telephone service to remote locations, such as oil fields, that could not be wired for telephone service. It did this by transforming analog voice signals received from the telephone headset into radio signals and vice versa. Between 1959 and 1966 approximately thirty-five hundred Carterfones were sold throughout the United States and abroad (Brotman 1987, 184).

AT&T attempted to prohibit the connection of the Carterfone and other devices to its network, through the tariffs that it filed with the FCC and state regulators. Basically, a tariff is a published schedule of rates for services; but it may also contain a list of obligations and restrictions that apply to the use of the service. In the wake of the *Hush-A-Phone* decision, AT&T modified its tariff to permit the use of rudimentary devices such as the Hush-A-Phone. But the revised tariff introduced an explicit prohibition on the use of any device that involved *a direct electrical connection* to its network, or connected its lines with any other communications device (Brock 1981, 240). Referring to the conditions outlined in the revised tariff, AT&T advised its subscribers that the use of the Carterfone was in violation of the tariff. A subscriber who continued to use the device would be subject to the penalties prescribed in the tariff.

Carter brought a private antitrust suit against AT&T. The courts ruled that it was within the jurisdiction and the expertise of the FCC to judge the fairness of the tariff. In October 1966 the commission announced that it would be holding a public hearing on the question of the AT&T tariffs. In 1968 the FCC decided in favor of the plaintiff, the Carterfone company. In its ruling the commission stated:

We hold that the tariff is unreasonable in that it prohibits the use of interconnecting devices which do not adversely affect the telephone system . . . The principle of Hush-A-Phone is directly applicable here, there being no material distinction between foreign attachment such as Hush-A-Phone and an interconnection device such as the Carterfone, so far as the present problem is concerned. Even if not compelled by the Hush-A-Phone decision, our conclusion here is that a customer desiring to use an interconnecting device to improve the utility to him of both the telephone system and a private radio system should be able to do so, so long as the interconnection does not adversely affect the telephone company's operations of the telephone system's utility for others. A tariff which prevents this is unreasonable; it is also unduly discriminatory when, as here, the telephone company's own interconnecting equipment is approved for use. The vice of the present tariff, here as in Hush-A-Phone, is that it prohibits the use of harmless as well as harmful devices. (FCC 1968, cited in Brotman 1987, 186)

In making its case, AT&T argued that the issue was not limited to the question of whether the device would cause harm to the network. The company contended that even a harmless device could prevent the company from fulfilling its system responsibilities since it would divide responsibility for ensuring that all the parts of the network functioned properly. Moreover, it argued that the attachment of independent equipment would retard system innovation, since independent suppliers would likely oppose improvements involving changes to network specifications that would result in costly modifications to equipment. These were specious arguments that the FCC rejected out of hand.

The *Carterfone* decision reaffirmed the principle established in *Hush-A-Phone* and extended its application to terminal equipment involving a more direct connection to the network and a more immediate substitution for AT&T equipment. Although it widened the breach in AT&T's protected monopoly, it did not produce widespread competition in the terminal equipment market.

In the wake of *Carterfone*, AT&T introduced a set of terminal connection requirements that were ostensibly designed to protect the network, but were really designed to stave off competition. These measures were referred to as "protective connecting arrangements" and were incorporated as part of new tariffs filed before the FCC. Simple devices such as Hush-A-Phone could be used in conjunction with Bell System equipment. Terminal equipment such as private branch exchanges (PBXs), computers, and private microwave systems that involved an electrical connection could only be connected through an intermediate device, an interface, that AT&T would provide as a service to the subscriber on a tariffed basis. Two kinds of devices were introduced by AT&T: Data Access Arrangements (DAA), for data communication, and Connecting Arrangements (CA) for phone use. All terminal equipment that entailed signaling communication with the network, such as the standard telephone apparatus, had to be provided by the telephone company (Kellog, Thorne, and Huber 1992, 504–505).

The effect of the connecting arrangements was to make it unprofitable to supply most kinds of alternative terminal equipment. However, there was some growth in the market for independently supplied terminal equipment. Brock notes that by 1974 independent suppliers were receiving revenues equivalent to $96.7 million or 3.7 percent of the PBX and key telephone system (KTS) market in areas served by the Bell System (Brock 1981, 244–245). PBX is a switching system designed for medium to large business. It is usually located on the company's premises and provides a business with several hundred to several thousand telephone extensions. KTS was a multi-button telephone device that was able to receive several outside telephone lines, thereby allowing several users to use more than one line. From the user's point of view the principal difference between a PBX and KTS was that users of a PBX had to dial "nine" before they could use an outside line.

The FCC did not explicitly approve the new AT&T tariffs based on the connecting arrangements. Nor did it invalidate them. Instead it referred the matter back to its Common Carrier Bureau for a negotiated approach that would bring together the telephone companies, competitive suppliers of equipment, and other interested parties (Kellog, Thorne, and Huber 1992, 505). These negotiations dragged on for seven years. Finally, it was determined, after consultation with agencies such as the National Academy of Sciences, that it would be possible to protect the quality and the integrity of the public network by establishing a program for the certification and registration of terminal equipment. In October 1975 the FCC authorized such a program, preempting any attempt at the state level to restrict terminal connection. The new certification program would apply equally to the telephone companies and the independent suppliers of equipment. In adopting this program, the FCC was abandoning its previous approach base on a futile attempt to make definitional distinctions between unauthorized devices that substituted for basic telephone company equipment, and equipment that *complemented* telephone company equipment, such as the Carterfone, which the FCC had sanctioned (506). Thereafter, the only pertinent question for the regulator would be the question of harm to the network. The subscriber had the right to use equipment provided by independent suppliers, including equipment that substituted for telephone system equipment.

Predictably, AT&T was able to forestall vigorous competition by making an appeal to the courts. During the appeal process a stay on the implementation of the certification program was in effect. When the appeals court sided with the FCC, AT&T announced its intention to go to the Supreme Court. This action prolonged the stay on implementation. When in October 1977 the Supreme Court refused to hear the appeal, the certification program should have gone into effect. However, AT&T then played its last card which it referred to as the "primary instrument concept." The FCC rejected this approach in July 1978 (Brock 1981, 248). The way was now prepared for residential subscribers to purchase their own telephone sets. In 1978 telephone sets began to become widely available in retail stores throughout the United States.

It is a testament to AT&T's perseverance that even as the market was opening, the company was using the primary instrument concept to frustrate the growth of competition. At that time, it was routine for telephone companies to charge a monthly fee for each additional telephone line (extension) a subscriber used, over and above the primary telephone. Of course, under the pre-competitive regime this "extension" comprised both the extension line and the telephone apparatus provided by the Bell System. The FCC's certification program contained a requirement that subscribers notify the telephone company when they were using foreign equipment. Through a dubious stroke of cunning, AT&T then attempted to use this information to penalize subscribers for the use of the foreign telephone device. Instead of charging the subscriber directly for using a foreign device, it offered a partial rebate on the second "extension line" to subscribers that owned their own telephones. Ultimately, this scheme was doomed to failure as it was impossible to enforce. Users simply ceased to declare these other telephones. For all intents and purposes, the market for terminal equipment, including residential telephones, was truly competitive by the early 1980s.

5.4 SOME TECHNICAL NOTES ON TECHNOLOGY AND SERVICES RELEVANT TO THE LONG DISTANCE MARKET

The market for long distance telecommunications is complex, comprising numerous services, directed at distinct clienteles, and delivered through diverse technologies. Before we begin our analysis of competition in long distance, it is necessary to briefly identify some of the technologies that were being developed, and the services that were being contested, during the period 1949 to 1982. Because regulation was an important protection against competition in long distance, we will also consider the legal status of AT&T's long distance monopoly. This will be done in section five. We will begin with a discussion of the technology. By introducing this material at this time we hope to preclude any technical digressions that would detract from the analysis of the central themes of the section on competition in long distance.

Copper wire has been the backbone of the telephone network since the introduction of the first commercial service in the late 1870s. More specifically, a pair of copper wires provides the basic connection between a local telephone circuit and a switching center, also referred to as a local exchange or central office. Because the copper wires are twisted around each other, they are often referred to as twisted pair copper wiring, or just twisted pair. Twisted pair technology is not ideal for transmission over long distances. Over distances of more than one mile, signals transmitted over twisted pair will fade and are subject to background noise. A more suitable wire technology for transmission over long distances is coaxial cable.

A coaxial cable consists of a conductor surrounded by another conductor. The outer conductor, which is a wire tube, protects the inner conductor from background noise and prevents the signal from escaping. When amplifiers, called repeaters, are spaced along the path of a coaxial cable it is suitable for long distance telephony. Coaxial cable is a terrestrial technology. As such it must be strung from poles or laid underground, usually along established rights-of-way, such as roads, highways, and railroads, and can be very expensive. Until the invention of microwave technology, long distance telephony was almost totally dependent on this terrestrial technology.

In 1939 American and British scientists who were working on improvements to radar technology as part of the war effort, developed a new kind of vacuum tube called the magnetron. The magnetron generated short radio waves at frequencies above 1,000 megahertz (MHz). These waves were ideally suited for radar as they could penetrate rain, smoke, and fog. Moreover, they could also be used to carry point-to-point analog voice and television (TV) signals. By placing microwave relay towers at twenty-five mile intervals it was possible to create a long distance network that was more economic than the terrestrial coaxial network. The first experimental microwave network for the transmission of TV signals was implemented in 1944 between Boston and New York. By 1948 a commercial service between the two cities was in operation (Meyers 1989, 178).

The early 1960s saw the introduction of the first communication satellites. Following the successful launching of the *Sputnik 1* satellite (USSR) in 1957, and *Explorer 1* (United States) in 1958, a number of communication satellites were successfully placed into orbit around the earth. The first geostationary communications satellite, *Early Bird* (also known as *Intelsat 1*), was successfully launched in 1965. On May 2, 1965, the first transatlantic, two-way TV broadcast occurred using *Early Bird*. A geostationary satellite lies in an orbit 22,300 miles (35,900 kilometers) above the earth. At this high altitude, the satellite travels around the earth at the same rate as the earth rotates on its axis. This produces a geostationary result: the satellite is permanently stationed above a point on the earth's surface. In many respects a geostationary satellite may be thought of as a very tall microwave tower. In 1993 there were approximately 150 satellites in geostationary orbit around the earth (Pecar 1993, 35).

A more recent transmission technology is optical fiber. The invention of the laser in 1960 opened the way for message transmission using pulses of light. In 1977 the first fiber-optic commercial system was introduced (Meyers 1989, 178). Optical fibers are highly transparent strands of glass or plastic surrounded by cladding. The hair-thin fibers are an excellent medium for transmitting pulses of light. The cladding that surrounds the strands is designed to reflect light inward, thus preserving the light impulse. A message that begins as an electric signal is transformed by a laser into pulses of light. These pulses are carried through the fiber-optic cable to a light sensitive detector at the other end of the cable that transforms them back into electrical impulses. In this manner communication originating as voice, data, or video images can be encoded digitally for transmission as light signals. An optical fiber can carry signals over a hundred miles without distortion. Today, fiber-optic cable is the technology of choice for many long distance applications. Although the use of fiber-optic cable has increased exponentially since the late 1970s, this technology came too late to be a major factor in the early liberalization of markets for long distance. Microwave and satellite technology played a more important role.

Twisted pair, coaxial cable, and fiber-optic cable are examples of *bounded* transmission media. By this we mean that they require a physical link between two points in order to direct a signal along a communications channel. Broadcast radio and TV, microwave, and satellite communication systems are based on the propagation of electromagnetic waves. These media are *unbounded*. They do not require a physical connection; rather they rely on transmission through space or air. As a result, unbounded media can be used for both point-to-point communication, and point-multipoint communication (common to broadcasting). Bounded media are not ideal for point-multipoint communication. On the other hand, signals transmitted over unbounded media are frequently subject to interference. One of the advantages of bounded transmission media is that it is easier to protect the signal from environmental noise. Indeed, it is this characteristic of coaxial cable that explains the success of cable TV. During the early years of cable TV, before the explosion of programming, its principal benefit was its ability to improve the quality of the TV signal. In effect, a bounded media was used to transform a point-multipoint signal, propagated through the airwaves, into a land-based system. Another common example of interference problems is the cordless telephone, which on occasion may pick up conversations from a neighbor's telephone. As mentioned previously, one of the disadvantages of bounded media is that a significant investment in physical plant is required to run the lines along poles or underground.

The choice of media for point-to-point communication, whether it be bounded (twisted pair, coaxial cable, or optical fiber) or unbounded (microwave and satellite) will depend on previously noted factors of reliability and cost. It will also depend on the characteristics of the communication signal to be transmitted. In other words, choice of media will depend on suitability for different kinds of use. Essentially, there are four kinds of communication signals: program, speech, video, and data. Radio broadcasts and the audio portions of TV broadcasts are examples of program signals. Speech is the most common type of communication over the telephone network. Video signals are the basis for the transmission of TV programming. Data signals are streams of binary digits (0s and 1s) suitable for machine communication between computer devices (Meyers 1989, 122–123).

Each of these signal types has distinct transmission requirements. Speech may be transmitted in analog mode in a bandwidth range between 200 and 3,500 hertz (Hz). Program signals require bandwidth in analog mode in the range between 50 and 15,000 Hz. Video signals require bandwidth in the range of 6 MHz (6,000,000 Hz). A hertz is a measure of the signal amplitude (high and low variations of an electrical wave) over a set period of time. One hertz is equivalent to one cycle per second. One thousand cycles per second is equivalent to 1,000 Hz or 1 kilohertz (kHz). Bandwidth is simply the *range* of frequencies (i.e., cycles per second) that a given medium is able to carry. Program signals, speech, and video can be transmitted using electrical signals that are *analogs* for the original signal (table 5.1). An analog electrical signal will vary *in direct proportion* to changes in the original signal. For example, variations in loudness of speech will be reflected in proportional (analogous) changes in the amplitude of the electrical signal.

Analogous transmission does not require that changes in the signal be registered in terms of discrete values. That is to say, changes to (variations in) an analog signal occur along a continuous (but limited) field of possible values. This is radically different from data communication where a signal can only

Table 5.1   Bandwidth Requirements for Different Types of Signals

Speech	200–3,500 Hz
Radio	50–15,000 Hz
Video	6,000,000 Hz (6 MHz)

have one of two values: zero or one. Moreover, the requirements for data transmission are not measured in terms of frequency (i.e., cycles per second of changes in amplitude). Rather, data transmission is measured in terms of bits per second: the number of zeros or ones that can be sent from point A to point B in one second. This is also referred to as "throughput." It is possible, however, to transform analog voice, audio, and video signals into digital form for digital transmission. This is done through encoding and decoding devices at each end of the channel. These devices convert analog into digital (at the sending end) and digital back into analog (at the receiving end) by sampling analog values and quantizing them, that is, giving them numerical values. For example, changes in the level of amplitude will be broken down into a discrete set of decimal values and then encoded into their equivalent binary number for transmission. More specifically, the digital transmission of analog voice involves a sampling technique that separates amplitude levels into 128 discrete segments (Meyers 1989, 118). The digital capacity (or throughput) of a channel is a function of bandwidth: the higher the allowable bandwidth, the greater the number of bits that can be transmitted per second.

The growth of computer data networks increased demand for networks for digital transmission. The explosion of data communication, coupled with increased demand for intercity video transmission spawned competition between existing transmission media (twisted pair and coaxial cable) and emerging media (microwave and satellite). Demand for new services, coupled with the invention of new technology capable of meeting this demand, created ideal conditions for competitive entry into long distance markets. Understanding the capabilities and limitations of the different transmission media is a key to understanding the relevant strengths and weaknesses of the Bell System as it faced the threat of increased competition. But technology is not the whole story; it is also necessary to understand how AT&T and its competitors transformed technology into service packages designed to meet the specific demands of different classes of users.

Common carriers of long distance telecommunications services cater to two distinct markets or clienteles: a market for *public network services* and a market for *private network services* (not to be confused with privately owned installations). The distinction between public and private networks is not based primarily on the technologies employed. Although the architecture (design) of public and private networks is different, the technologies they employ are, for the most part, the same. What differentiates them are the needs (requirements) of the respective clienteles, and the manner in which they are packaged as services by common carriers to meet these needs.

Private networks, based on circuits that are leased from common carriers, are designed to satisfy the needs of large institutions such as corporations or governments who are heavy users of telecommunications services. Heavy use is a function of large, geographically dispersed organizations. An organization with a head office in one major city and branch offices in several other cities throughout the country, or indeed throughout the world, will have high volumes of communication between the different locations of this closed organizational system. Moreover, the types and volumes of communication activity are known by the organization and are predictable. Because communication occurs at high volumes between a fixed number of points, the telephone companies are able to offer bulk discounts to these institutions. Typically, these high capacity circuits are capable of multiple voice transmission over a single channel or high-speed data communication. For example, the Bell System introduced T1 carrier service in 1962; it is capable of carrying twenty-four simultaneous voice conversations (Pecar 1993, 41). In order to do this, T1 service combines two pairs of twisted pair wire (four wires) with multiplexing technology at each end to augment the signal capacity of the circuit. Companies lease T1 circuits on a monthly basis. T1 circuits are suitable for making local connections between a client's premises and the telephone company central office. For longer distances it is possible to lease a fixed number of circuits on a microwave or satellite channel.

Although technology is an important component of private networks, it is not the defining characteristic. More important to the distinction between public and private networks is the formula that is employed for billing clients. In a private network, the client leases a fixed number of circuits from a common carrier on a *monthly basis for a fixed rate*. The client/subscriber has exclusive use of the circuit. Whether the circuits are used frequently, infrequently, or not at all, does not affect the price of the service, which is determined by the monthly rate. This service is radically different from the way users are billed for the use of the public long distance network. Metered billing is the norm for public long distance service. Under metered billing, charges are levied on the basis of *minutes of use*. It goes without saying that a dedicated long distance line provided by a common carrier cannot be used to make direct connections to the public network. If this were the case it would be possible to use the discounted leased-line to bypass the more expensive public network.

For large organizations, which are heavy users of telecommunications, the use of metered billing for long distance communications can be very expensive. In response to the specialized, internal communication needs of these organizations, common carriers provide private circuits to corporations at a *discount rate*. Also referred to as "leased-lines," "private lines," or "dedicated lines," what distinguishes this type of service from the public network is: (1) the fixed nature of the connections and the essentially closed nature of the network, and (2) the billing mode, which is based on a fixed monthly rate. Although the circuits are provided on an exclusive basis to a specific client, the rates themselves are set as part of a *public* tariff filed by the common carrier before a regulatory commission. The same rates apply for equivalent service to all users of private lines provided by a common carrier.

It warrants mention that heavy users of telecommunications services are not obliged to lease lines from common carriers. A corporation may decide to own and operate its own network for organizational communications. For most companies it is more economical to lease dedicated lines from a common carrier. However, for companies that already control rights-of-way, such as railroads, gas, and electricity companies, owning and operating a network may represent an efficient solution to the problem of reducing the costs of long distance communication.

5.5 AT&T'S LONG DISTANCE MONOPOLY: A MATTER OF REGULATORY INTERPRETATION AND PRECEDENT

AT&T's monopoly was de facto, rather than legal. Nowhere in the Communications Act of 1934 was it prescribed that long distance telephony service should be offered by AT&T, or any other company, on a monopoly basis. However, as Michael K. Kellog, John Thorne, and Peter W. Huber have noted, the Communications Act of 1934 did contain an "unstated but clear premise . . . that telephone service is best provided by way of a monopoly" (1992, 149). They draw this conclusion from the fact that the act required all firms operating in the interstate market to receive prior authorization from the commission before constructing or extending lines. Approval would be contingent on a company's ability to demonstrate that "the public convenience and necessity" required it (section 214(a)). This, coupled with the fact that the interconnection of networks was subject to the commission finding "such action necessary or desirable in the public interest" (section 201(a)), placed a heavy burden of proof on firms wishing to establish themselves in competition with incumbent common carriers. AT&T's de facto monopoly was also the result of decades of informal negotiations with the FCC, punctuated by periodic interventions from the DOJ. Under the Communications Act of 1934, the power to encourage or suppress competition resided with the FCC. As we have seen, the FCC had the power to encourage competition by ordering AT&T to allow interconnection with its network. Alternatively, it could throttle competition by refusing applications for interconnection.

Since the creation of the FCC in 1934, the cornerstone of the commission's relationship with AT&T had been an implicit understanding between the two parties that competition and price regulation were incompatible. AT&T held that it would be unfair to sanction competition with AT&T so long as the company's prices and revenues were set by the regulator outside of the market. If entry into telephone markets were to occur while AT&T's prices were being regulated, the result would be one-sided competition, which would not be competition at all. This would be the case since regulation would hamper AT&T's ability to respond. In other words, "price" and "entry" regulation went hand in hand. AT&T's protection from competitive entry was grounded in this implicit understanding between the regulator and the regulated. But this protection was not enshrined in legislation; it was a function of regulatory tradition. By the same token, the FCC was not bound by anything except conventional wisdom, and its own precedents, to discourage competitive entry. Clearly, AT&T expected that it would continue to be protected from competition in the long distance market as a result of the 1956 Consent Decree. If not, it is doubtful that the company would have agreed to exclude itself from activities in all other markets.

In the years prior to the first Consent Decree (1956), the FCC was able to negotiate a series of rate reductions for interstate, long distance service. These reductions occurred under the board-to-board approach to cost distribution for telephone service. The term "board-to-board" was shorthand for switchboard-to-switchboard. The board-to-board approach was the outcome of a court decision in 1913 that ruled that costs for long distance and local service should be fully segregated (Horwitz 1989, 103). Under this approach, the costs of long distance were to be fully recovered from fees for long distance; in a similar fashion, the costs of local operations were to be fully recovered from local service fees. The board-to-board approach turned a blind eye to any interdependence between the two levels of network services. Understandably, the individual states, which were responsible for regulating rates for local service, favored an alternative approach, one that recognized the interdependence of local and long distance facilities. There was considerable justification for this since without local service one could neither initiate nor terminate a public long distance call. They urged a station-to-station (i.e., telephone station to telephone station) approach that acknowledged the contribution of local networks to both the initiation and the completion of long distance calls.

Under pressure from the states, the FCC began to reverse its board-to-board approach and allow local network costs to be transferred to long distance operations. Harry M. Trebing argues that this was "the first error in FCC policy that would eventually undermine the credibility of natural monopoly" (1989, 96). It is clear, in retrospect, that the FCC's timing could not have been worse. At the very moment the agency was acquiescing to the demands of the state regulators to transfer local service costs to long distance, the actual cost of long distance was declining as the result of microwave, and eventually, satellite technology. The ensuing imbalance between long distance costs, which were on the way down, and prices, which rose or remained unchanged, undermined the credibility of monopoly regulation. The principal beneficiaries of the transfers (subsidies) were residential users.

Although it may have been politically astute for the FCC to address the needs of this constituency, the sanctioning of cross-subsidies had the effect of linking FCC policy more closely than ever to the anticompetitive reflex of the Bell System. In fairness, how could the agency open markets to important segments of AT&T's core business, such as long distance, while at the same time soliciting AT&T's cooperation in a policy of revenue transfers that had no intrinsic economic benefit for AT&T, and which had the potential to reduce AT&T's aggregate revenues should competitive entry prove to be successful. As long as the cross-subsidies were in place, the regulator had a vested interest in protecting the monopolist. In fairness to the FCC, its mandate, as stipulated in the Communications Act, was to promote a universal telephone service:

> For many years the paramount regulatory objective in all of telephony, at both state and federal levels, was to promote universal service: a telephone within arm's reach of the chicken in every pot. The Communications Act established this as one of the FCC's several missions. The dominant objective, therefore, was to push local rates down. As for universal service to other telephone companies, equipment providers, sellers of long-distance service and such, well, the Communications Act hadn't meant *that* kind of universality. Universal service was something for consumers, not providers. (Kellog, Thorne, and Huber 1992, 21)

5.6 STEP-BY-STEP TOWARD COMPETITION IN LONG DISTANCE

The first breach in AT&T's long distance monopoly occurred in the market for *privately owned,* point-to-point networks. In the late 1940s and early 1950s the FCC had authorized a limited number of private, point-to-point networks using microwave technology. The primary issues in these early licensing proceedings were the availability of service from AT&T and Western Union and the availability of radio frequencies. Under the Communications Act of 1934, the FCC was responsible for allocating radio frequencies including those used for microwave transmission. At the time, there was concern that the radio spectrum was scarce; thus requiring careful management by the FCC to ensure that frequencies were allocated wisely in the public interest. Other than the principal long distance carriers (AT&T and Western Union), the beneficiaries of these licenses were right-of-way companies (pipelines and railroads) and broadcasters who, at the time, were unable to receive adequate service from the common carriers. Broadcasters were constructing microwave installations to facilitate intercity transmission of TV programming. However, the FCC saw the allocation of microwave frequencies to broadcasters for their private use as a temporary measure. As they moved to deployed microwave installations for the intercity video market, it was assumed that the common carriers would be favored in new spectrum allocation.

But interest in privately owned networks was not limited to TV broadcasters and public utilities. A number of trade associations combined efforts with manufacturers of microwave equipment in order to lobby the FCC for a more liberal approach to the allocation of microwave frequencies, one that would favor the construction of private installations. These groups included the Automobile Manufacturers Association, the National Retail and Dry Goods Association, the American Newspaper Publishers Association, and the National Association of Manufacturers (Horwitz 1989, 225). The FCC undertook a review of its microwave policy at the end of 1956. In 1959 the commission announced a more liberal policy towards private use of microwave frequencies. Because the decision concerned the allocation of frequencies above 890 MHz, it became known as the *Above 890* decision (FCC 1959 and 1960).

In reaching its decision, the FCC considered two issues: (1) whether enough frequencies were available to meet the needs of both common carriers and private owners; (2) whether granting licenses to operate privately owned facilities would have an adverse economic effect on common carriers—one that would compromise their ability to meet their service obligations in the public telephone network. The commission was unable to find evidence supporting AT&T's claim that spectrum frequency was too scarce to allow assignment to private users. Concerning the economic impact of licensing private point-to-point systems, the commission noted that the common carriers had not presented evidence supporting their claims: "While the common carriers have claimed that there will be adverse economic effects, they did not disclose the detailed basis and extent of such effects so as to enable appropriate evaluation of their claim. Thus the record is inconclusive on the question as to the specific nature, extent, and magnitude of any detriment which the licencing of private point-to-point systems would have on the ability of common carriers to serve the general public" (FCC 1959 and 1960, quoted in Brotman 1987, 183).

The *Above 890* decision did not open the floodgates to competition in long distance. First and foremost, the decision was limited to privately owned microwave facilities. Consequently, it did not have a bearing on AT&T's principal line of business: public telephony. Second, the option of creating a private network using microwave technology was simply too costly for most potential users. Finally, AT&T was able to respond by lowering rates for the types of bulk services that were the impetus for the creation of private networks in the first place.

Although the decision had a limited immediate impact on AT&T's monopoly, it was an important step in the direction of market liberalization. The decision was notable because it cleared the air with respect to a fundamental question of whether common carriers had a right to blanket protection from competition by virtue of their public service obligations. The *Above 890* decision made it clear that the obligation to provide service to the public wherever it was economically feasible, and at a reasonable price, did not immunize the common carrier from competition in markets adjacent to its core public business. The decision effectively began the policy process of separating the market for long distance services into distinct market segments. For the purpose of future policy, it would now be possible to disassociate AT&T's principal activity, the operation of the public telephone network, from other segments of the market for network services. This did not mean that the FCC was indifferent to the plight of the common carrier. The commission would continue to monitor whether revenue losses resulting from alternative networks were having an impact on universal service. But following the *Above 890* decision, it was clear that the *burden of proof* now rested with the common carrier to demonstrate how competition would undermine its core business. This was an important shift in policy.

The *Above 890* decision was significant in another respect: AT&T's response. Regulation was AT&T's first line of defense against competition in the long distance arena. When it failed, the company responded by introducing a new set of discount tariffs for the large volume, corporate market. Known as Telpak, the new tariffs, which were introduced in 1960, were designed to stifle the emerging market for privately owned networks. There were four Telpak tariffs, each reflecting a different package of private lines: 12 lines (Telpak A), 24 lines (Telpak B), 60 lines (Telpak C), and 240 lines (Telpak D). The amount of the discount increased with the number of lines leased. It varied from 51 percent for 12 lines, to 85 percent for 240 lines. The amount of the discount was particularly significant at the high end of the scale. Previous to Telpak, a user with 240 lines could expect to pay AT&T $75,600 a month (Brock 1981, 207). Under the new rates the same user could expect to pay $11,700 per month. The Telpak tariffs were designed to have their greatest impact where privately owned microwave installations were the most economical: large organizations with heavy volumes of communications traffic. For organizations whose traffic volumes were too low to justify investment in a private network, the discount was lower. Of course, this raised the question of the relationship between cost and price. Were the discounts higher for Telpak D service because of economies of scale related to the provision of a large number of circuits? In other words, was there a cost-price relationship at the high end of the Telpak scale? Or were the discounts simply an anti-competitive response to potential competition, a price response that had nothing to do with lower operating costs? If this were the case, then AT&T ran the risk of being accused of predatory pricing, an antitrust violation.

Predictably, an investigation was launched by the FCC into the lawfulness of the Telpak rates. This was done after Motorola, an equipment manufacturer, filed a complaint with the FCC. The proceedings dragged on for four years before the commission reached a decision in 1964. The general finding of the commission was not favorable to AT&T and its discounted rates. They constituted, in the opinion of the commission, a form of price discrimination "without cost or service differences to justify it" (Brock 1981, 208). These rates were illegal under the Communications Act. However, if it could be demonstrated that the rates for discounted services at least covered their costs, and that the discount was a justifiable response to competition, they could be allowed to stand. Because Telpak A and Telpak B were clearly not a response to competitive conditions, they were immediately thrown out. The FCC spent the next twelve years trying to determine whether the price of the remaining discounted services covered their costs. The excessive delay was due to the fact that the Telpak proceedings became part of a more comprehensive FCC examination of AT&T's rates, and the methods used to calculate them, known as Docket 18128 (Horwitz 1989, 226). In 1976 the FCC ruled that the Telpak tariffs were unlawful.

The *Telpak* case bears at least one similarity to the *Carterfone* case: the protracted nature of the regulatory proceedings enabled AT&T to stave off competition for over fifteen years. Even when the company lost its case before the FCC, it was able to use regulatory due process to forestall change. One of the ironies of the Telpak saga was the manner in which corporate users of the Telpak service rallied in support of AT&T. Recall, that these were the same corporations that supported a more liberal assignment of frequencies above 890 MHz. But when AT&T introduced its discount package, many of them lost interest in owning and operating their own networks. Not only did AT&T cripple competition with its discount package, but it earned the support of a powerful lobby in subsequent proceedings before the FCC. Trebing has argued that the *Above 890* decision to liberalize the use of microwave facilities should have been followed by a ruling establishing price guidelines for common carriers. According to Trebing, this was the commission's second major error. In the absence of a comprehensive approach, the commission was forced to deal with AT&T's new discounted tariffs on a case-by-case basis. The credibility of monopoly regulation was undermined when AT&T was able to frustrate the FCC's more liberal approach to private networks (1989, 97).

The high cost of private microwave installations dictated that these systems were only suitable for large organizations. Moreover, as we have demonstrated, AT&T was able to meet the needs of many of these organizations through its Telpak C and Telpak D services. But the needs of small- and medium-sized businesses were not being served. Also, companies with specialized communication needs (such as computer service companies requiring high-speed data transmission) complained that the common carriers were not providing adequate service. Because these needs were not being met by the incumbent common carriers, an opportunity was created for a new type of service: a limited common carrier microwave service for the interoffice needs of small business.

In 1963 Microwave Communications Incorporated (MCI) petitioned the FCC to authorize construction of what amounted to a *shared* private network. Today, MCI is the second largest long distance carrier in the United States. At the time of the application, however, it was an inexperienced, underfinanced entrant to the telecommunications business. MCI proposed to offer a discount microwave service between Chicago and Saint Louis with inter-mediate service available at nine points between the two cities. The service was limited to transmission between MCI's microwave installations. Loop service between MCI's installations and the subscriber's premise had to be provided by the subscriber. MCI's service package was flexible: allowing up to five subscribers to share a voice channel. Subscribers could lease channels for the entire length of the network or parts of it. Broadband users could lease channels in increments of 48, 250, and 1,000 kHz. During the evening hours, channels used for voice communication would be combined to offer addi-tional broadband service for high-speed data transmission.

Although specialized and limited, what MCI was proposing was clearly a com-mon carrier service. As could be expected, the principal long distance carriers (AT&T, Western Union, and General Telephone) rallied to oppose the applica-tion. They argued, with considerable justification, that MCI was cream skim-ming: offering service in the most lucrative part of the market without fulfilling any of the costly obligations born by the other common carriers. Moreover, they contended that without system-wide price averaging, a policy endorsed by the FCC in order to promote universality, there would be little or no incentive for MCI to enter the market for specialized long distance service. It was the policy of price averaging that artificially inflated prices for service on popular routes. If the incumbent common carriers were forced to compete with MCI for this mar-ket, they would be forced to abandon price averaging, and rates for lightly used long distance routes would go up. The incumbents were on less secure ground when they argued that there was no demonstrable need for the service, and when they suggested that MCI lacked the technical expertise and financial re-sources to operate such a network.

When the FCC ruled in 1969, it was a very close decision: 4-3. The commis-sion discounted the incumbents' claim that entry would jeopardize price averag-ing on the grounds that MCI's proposed service was too limited and too special-ized to pose a threat to universality. On the question of inefficient duplication of services, the commission agreed with the hearing examiner, Herbert Sharfman, that "MCI's lower rates, and more flexible use would enable it to serve a market whose needs are unfulfilled by the available common carrier services" (Brotman 1987, 191). In effect, with a very narrow margin of support among the seven FCC commissioners, a decision was made to allow a little competition at the fringes of the long distance market. No doubt the commissioners who favored competitive entry, albeit at the fringes of common carrier service, believed that they would be able to manage subsequent events. As we shall see, they were mis-taken. The FCC would soon lose control of long distance policy when the case landed in the courts. MCI finally opened for business in January 1972, almost ten years after its initial application before the FCC. It had cost the company $10 million in legal and regulatory costs to advance its application through the FCC. The actual cost of the facilities it built was under $2 million (Brock 1981, 213).

In the wake of the *MCI* decision the FCC received numerous applications from MCI and other companies who wanted to offer private-line service in other regions of the country. Rather than dealing with each application individually, the commission elected to establish a general policy to deal with entry into the "specialized" common carrier market. Under the *Specialized Common Carrier* decision of 1970 (FCC 1970b), a more liberal regime for licensing private-line services was implemented. Brock notes that by the end of 1973 MCI had established an $80 million transcontinental network serving forty U.S. cities. Southern Pacific Communications, a competitor to MCI, also established a coast-to-coast network (Brock 1981, 215). Recall that under the first *MCI* decision it was up to subscribers to secure a local connection between MCI and their premises. AT&T's subsidiary, local operating companies, were best equipped to provide these local dedicated circuits. In the *Specialized Common Carrier* decision the FCC anticipated the need for local connections. It expected the established carriers to provide these circuits upon request and on reasonable terms. Predictably, the Bell System companies resisted. This time they did so by filing tariffs with the state regulatory boards, arguing that local connections were an *intra*state matter and, therefore, outside the FCC's jurisdiction. AT&T also initiated a price response to private-line competition. In specific areas with heavy demand the company introduced a new discount tariff known as Series 11,000 (Brock 1981, 218). Of course, these were the same routes that MCI and the other specialized carriers had identified for their service. We will not recount the saga of legal and procedural delays that AT&T employed in an attempt to frustrate the FCC. It is sufficient to note that the battle for full interconnection for private-line services was finally settled by the courts in the FCC's favor, five years after the *Specialized Common Carrier* decision.

It is easy to criticize AT&T's response to competition as excessive and heavy-handed, very much like trying to kill a fly with a sledge hammer. But the record of subsequent action by the specialized common carriers, and by MCI in particular, suggests that AT&T's reaction may have been appropriate. In 1974, only two years after the successful opening of its private-line service, MCI moved to expand the breach in AT&T's long distance monopoly. Trebing notes that it was compelled to do so by the heavy financial losses it suffered in 1971 and 1972 (1989, 97). MCI filed a revised tariff before the FCC for an expanded class of service that it called Execunet. An Execunet customer could dial a local MCI number and through this number connect with any telephone in a distant city served by MCI. Billing would be done on a metered (i.e., per minute) basis. From MCI's point of view all it was doing was offering a new, more flexible option to its subscribers for the use of its "private" networks. Rather than sharing a circuit with five other subscribers, a user could now use a circuit on a metered basis. Unlike the initial offering approved by the FCC, where the subscriber's local connections were made via a dedicated line, MCI's new service was based on "dial-up" access to MCI installations. Anyone with a telephone in a city served by MCI would have access. In order to complete calls, MCI required access to local exchange facilities comparable to that provided to AT&T's long lines service.

Execunet was actually an expanded version of a private-line service that MCI was already providing to its customers known as foreign exchange (FX) service. FX service allows a subscriber to connect a telephone or PBX directly to a telephone exchange in a distant city using private lines, thus allowing a customer to make and receive calls as if he or she were located in the distant city. From MCI's perspective, Execunet was merely an expanded, more accessible version of FX service. Clearly, MCI was proposing to offer a *public telephone service*, albeit on a limited basis. The service would only be available for making and receiving calls in cities served by MCI. Because additional digits had to be dialed to initiate calls, the service was less convenient than that provided by the Bell System. Although the service would not be universal, it would be public. Under the umbrella of the FCC's favorable ruling on private-line services, MCI was proposing to get into the business of switched, long distance telephony. This was clearly not what the FCC intended when it had ruled in the *Specialized Common Carrier* decision.

The *Specialized Common Carrier* decision was essentially about the allocation of microwave frequencies. Although necessary, interconnection for local distribution was secondary. Execunet made *interconnection* (access) to the local Bell System network the central policy issue. On the one hand, the FCC had the power to order the Bell System companies to provide access to local exchange facilities. On the other hand, the *Specialized Common Carrier* decision clearly did not envision interconnection for the purpose of providing a dial-up public service, and there were no precedents in existing policy supporting the specialized common carrier's request for this type of interconnection. When AT&T protested MCI's revised tariff, the FCC responded expeditiously without even convoking a meeting of the commissioners. According to the FCC, the *Specialized Common Carrier* decision only authorized private-line communications, and Execunet was essentially a switched public telephone service (Kellog, Thorne, and Huber 1992, 603). MCI appealed to the courts arguing that the FCC should have initiated a proper hearing and considered the merits of the case. The court agreed. After receiving comments, the FCC reaffirmed its earlier ruling in June 1976.

Predictably, MCI appealed the FCC decision to the courts. Unpredictably, it won. In what could only be described as a bizarre twist of fate, the case was decided on a narrow technical point that did not take into consideration basic policy questions such as the impact of competition on universality or whether the new service was an example of cream skimming. The Communications Act required that the FCC issue construction permits for each microwave installation used in a private network. The District of Columbia Court of Appeals noted that the construction permits that had been issued to MCI did not contain an explicit limitation on the types of services that could be provided via the microwave facilities. Although the FCC argued before the court that the *Specialized Common Carrier* decision contained an *implicit limitation*, restricting the use of the facilities to private networks, the court was not convinced by this argument. This ruling by the court, issued in July 1977, became known as *Execunet I* (*MCI v. FCC*, 561 F.2d to 365 D.C. Cir. 1977).

AT&T responded to *Execunet I* by announcing that it would stop providing additional connections to MCI, and by returning to the FCC for a ruling on whether it was obliged to provide additional connections to MCI. In another accelerated proceeding the FCC again sided with AT&T in a decision rendered in February 1978. MCI appealed the FCC decision to the court. In its second *Execunet* decision the court again ruled in favor of MCI. Essentially the court reiterated its earlier position. The FCC and AT&T had built their case against expanded interconnection on the grounds that the initial *Specialized Common Carrier* decision envisaged limited interconnection for the exclusive purpose of providing private-line service. Again the court found that *Specialized Common Carrier* decision did not contain an explicit limitation on interconnection. If MCI wanted to provide an expanded service, it could do so by filing an appropriate revision to its existing tariff. If AT&T wanted to oppose MCI's request for interconnection it should do so in the context of the hearing on the revised MCI tariff, and not on the basis of the earlier *Specialized Common Carrier* decision. The court stated: "our emphasis on tariffs and rate making as the exclusive means for future limitations on the specialized carriers' development clearly contemplated that the carriers would be free to expand their service offerings—and would be afforded the necessary interconnections—until and unless it was found that the public interest demanded otherwise" (decision quoted in Brotman 1987, 220). This became known as *Execunet II* (*MCI v. FCC*, 580 F.2d 590 D.C. Cir. 1978).

The ruling of the court meant that the burden of proof in interconnection cases had shifted to AT&T: it was now up to AT&T to demonstrate that interconnection would jeopardize its ability to provide universal service. The FCC still had the power to issue conditional licenses that would stand up in the courts. But this power applied to future licenses; it could not be applied retroactively to MCI and the other specialized common carriers. The court had ruled that they had licenses that allowed them to compete unconditionally wherever facilities had been authorized (Kellog, Thorne, and Huber 1992, 609). They could continue to provide dial-up service in competition with AT&T pending a comprehensive ruling by the FCC on market structure for long distance. But this would take several years. In the meantime MCI and the other specialized common carriers would continue to roll out their long distance alternative to AT&T. For all intents and purposes it was too late to turn back. Once investment had been made in new installations, and a critical mass of new customers had been reached, it was unlikely (if not impossible) that the FCC would be able to revert back to a closed policy on entry.

According to Trebing, the FCC committed its third major error in policy when it ruled in the *Specialized Common Carrier* decision (1989, 97). The decision was based on the mistaken premise that the commission would be able to maintain a rigid separation of telecommunications services. Rapid changes in technology, coupled with the interference of the courts, made it impossible to introduce "a little" competition on the fringes of the telecommunications market. The separation of telecommunications services into *monopoly* and *competitive* services created industry boundaries that were essentially artificial. Pressure from equipment suppliers, competitive service providers, users, and ultimately the courts, made it increasingly difficult to justify, and nearly impossible to preserve, these boundaries for any extended period of time.

It warrants noting that on at least two occasions the FCC acted in an uncharacteristically expeditious manner to protect AT&T's monopoly from the competitive incursions of MCI. This contrasts with the torpid pace of deliberations when competitors sought relief before the FCC from AT&T's anticompetitive Telpak tariffs. On the other hand, in this case, the judicial process actually accelerated the introduction of competition rather than forestalling it. As a result of *Execunet II*, the interconnection rights of specialized carriers were firmly established; they became, so to speak, the legal norm. AT&T could continue to make appeals to the courts and the FCC, but while these appeals were running their course, the court clearly expected the company to provide access to its local networks. AT&T's long distance monopoly had been broken.

5.7 A LIBERALIZED MARKET FOR SATELLITE COMMUNICATIONS

Historically, the state has played a central role in the development of satellite technology, and in particular, in the development of satellite communication systems. This has been the case in the United States, in Canada, and elsewhere throughout the world. Unlike other transmission media, such as copper wire, coaxial cable, optical fiber, and microwave, the development of satellite communications has been advanced and coordinated to a considerable extent by the state as a matter of national policy. Satellite communication also requires a high degree of cooperation and coordination between nation-states. As a result, satellite communication systems, throughout the world, have been implemented as part of an explicit public policy process. This has been equally true of domestic satellite systems, designed for use within the boundaries of a single nation, and of international systems.

Numerous factors explain the high degree of state oversight and participation in satellite communications. First, it should be remembered that communication by satellite is actually a subset of potential uses of satellite technology. Satellites can be used for military purposes, for weather forecasting, as platforms for space exploration and astronomy, and so on. All these applications require launch technology if orbit is to be achieved above the earth. As a matter of national defense during the Cold War, the U.S. government heavily subsidized the development of missiles capable of launching satellites into space. During the early years of space exploration, companies did not launch orbital missiles, countries did. For the most part, this is still true today. Because of this investment, the government not only had a vested interest in how launch technology would be used, but also it was in a position to dictate the use. Second, the satellite communication requires the use of the radio spectrum, which has traditionally been managed and regulated by the state. Third, because satellite technology provides an effective means for communicating with remote areas of a country, it is ideally suited to regional development that may be a priority of the state. Finally, satellite communication is an effective means for overseas communication. This requires international cooperation, a role the agencies of the state are ideally suited to play.

In October 1958 the National Aeronautics and Space Administration (NASA) was established to administer the U.S. space technology program. In 1959 AT&T approached NASA with a proposal that would have ensured AT&T's domination of the satellite communication industry: AT&T wanted NASA to sanction AT&T's right to control the entire satellite communications field (Horwitz 1989, 150). Unfortunately for AT&T, the request was beyond NASA's jurisdiction since it was not a regulatory agency. Moreover, NASA was used to receiving competitive bids for contracts; the "monopoly" argument did not fit well with its usual way of doing business. Nevertheless, it was clear to the U.S. government that NASA and the FCC would have to work together if satellite communication was going to develop in an orderly fashion. In 1959 the Dwight D. Eisenhower administration ordered NASA to cooperate with the FCC. In 1961 NASA and the FCC reached an agreement on the division of responsibility in the satellite communication field. NASA would supervise and coordinate the technological phase of satellite communication. The FCC would determine the structure of the industry. In other words, the FCC would decide who would own the satellites and how the system would function.

In May 1961 President John F. Kennedy announced his government's intention to land a man on the moon. On the same occasion, he called for the United States to exercise its leadership in the emerging field of international satellite communication (Tunstall 1986, 66). Following the president's lead, the U.S. Congress acted quickly to pass the Communications Satellite Act of 1962, which reflected a compromise between the position taken by the FCC and AT&T, and that of Congress and the DOJ. The FCC, with the backing of AT&T, favored a closed industry structure that would limit ownership of *international* satellites to established international carriers, primarily AT&T and Western Union. The DOJ favored a competitive, free enterprise system open to communication carriers and equipment manufacturers (Horwitz 1989, 151). It thus hoped to prevent the type of the anti-competitive behavior in satellite communications that had typified AT&T's relationship to other telecommunications markets. Congress favored a free enterprise approach. The compromise, which was supported by the executive branch, created the Communications Satellite Corporation (COMSAT), a private corporation that was subject to regulation. COMSAT would operate as *carrier's carrier* with 50 percent of its shares owned by existing common carriers and 50 percent owned by the public. Equipment manufacturers and other non-common carriers would have a stake in the corporation through their ownership of public shares. COMSAT's board of directors would be chosen by the common carriers, public shareholders, and the U.S. president. Its mandate was to represent the United States in international discussions on a future satellite communications system. In August 1964 delegates from eleven countries met in Washington, D.C., to sign an agreement creating an international satellite consortium, the International Telecommunications Satellite (Intelsat) Organization. By mid-1968, the number of signatories had grown to sixty-one (Akwule 1992, 59). From its creation until 1978, COMSAT effectively ran Intelsat due to the fact that COMSAT controlled 61 percent of its shares (Tunstall 1986, 69). Since 1978 the consortium has been run by its director general and his staff (Meyers 1989, 311).

Robert Britt Horwitz has argued that the Communications Satellite Act was a compromise which AT&T could accept:

> In the short term, the Communications Satellite Act maintained the traditional structures in telecommunications. The rates, profits, and areas of activity of space communication became regulated just like other regulated communication service. To protect the existing structure of American international carriers, COMSAT was not to provide service to end users. Rather, COMSAT became a carrier's carrier for international telecommunications traffic, and thus did not compete with existing carriers.
>
> AT&T was able to get into a major new communications field while protecting its ocean cables. The arrangement transferred any antitrust implications of AT&T's satellite involvement to the new COMSAT entity. (1989, 152)

Intelsat launched its first satellite, *Intelsat 1* (also known as *Early Bird*), into orbit in April 1965. Essentially, Intelsat's function was to own and operate the *space segment* of the satellite communications network,[1] while the land-based, receiving antennae were controlled by the respective countries. The ownership and operation of receiving antennae would depend on whether a country's telecommunications infrastructure was privately owned (the United States) or publicly owned (e.g., Europe's Post Telegraph and Telephone administrations). In the case of the United States, AT&T was the principal operator of U.S. receiving stations for overseas communication. Moreover, in 1972 AT&T controlled 29 percent of COMSAT shares and utilized 60 percent of its international circuits (Tunstall 1986, 72).

The use of satellites for domestic communication was shaped by a different series of events. In September 1965 American Broadcasting Corporation (ABC) applied to the FCC to operate its own communication satellite for the purpose of transmitting network programming from New York to Los Angeles and to its network affiliates across the country. ABC had calculated that this would be cheaper than relying on AT&T's private lines. Jeremy Tunstall argues that ABC and the other TV networks were also preoccupied with the problem of an emerging cable TV industry. By owning the nation's communication satellites they would be able to control the growth of cable TV (1986, 71).

The FCC responded to ABC's request by launching a number of satellite inquiries that dragged on for five years. Among other things, the FCC had to determine whether the Communications Act of 1934 recognized FCC jurisdiction to regulate communication via transmitters located in space. Ultimately, it was determined that the FCC did have jurisdiction over satellite communication. Also COMSAT, supported by AT&T, argued that the Communication Satellite Act of 1962 mandated that COMSAT alone should be responsible for owning and operating domestic communication satellites (Bruce, Cunard, and Director 1986, 263). The FCC rejected this argument.

In 1970 the Richard M. Nixon administration urged the FCC in a written memo to consider a liberal approach to entry into the domestic satellite business. In 1970 the FCC announced that it would pursue an "open skies" policy, which would become known as the *Domsat I* decision (FCC 1970a). For the first time, the FCC would consider applications from private parties, and not just established carriers or government organizations. Brock notes that the FCC received proposals from "AT&T, COMSAT, General Telephone and Electronics, Western Union, MCI, Western Tele-Communications, RCA, Hughes Aircraft and Lockheed . . ." (1981, 259). In 1972, in a decision known as *Domsat II*, the FCC further liberalized the domestic satellite market (FCC 1972a). Licensing procedures became more flexible. For the first time they included a more liberal approach to the licensing of the ground segment of satellite communication systems (Kellog, Thorne, and Huber 1992, 598).

The FCC decision to establish the satellite communications industry on a competitive basis was historic. It represented a major shift in policy from the FCC's traditional approach to common carriage that had been protective of AT&T and generally skeptical of the benefits of competition. In the following quote we see one of the first strong statements of the new policy that would favor competition wherever possible:

> We are further of the view that multiple entry is most likely to produce a fruitful demonstration of the extent to which the satellite technology may be used to provide existing and new specialized services more economically and efficiently than can be done by terrestrial facilities. . . . The presence of competitive sources of supply of specialized services, both among satellite system licensees

and between satellite and terrestrial systems, should encourage service and technical innovation and provide an impetus for efforts to minimize costs and charges to the public. (FCC 1972a, quoted in Brotman 1987, 195)

The commission's endorsement of competition was all the more striking for the fact that it was unclear what the new transmission capacity would be used for. The commission noted:

> Although the satellite technology appears to have great promise of immediate public benefit in the specialized communications market, here too there are uncertainties as to how effectively and readily satellite services can develop or penetrate the market. . . . To be sure, the applications generally do not identify specific services that are new or innovative. However, in our judgement, the uncertainties as to the nature and scope of the special markets and innovative services that might be stimulated will only be resolved by the experience with operational facilities. (194)

Although the open skies policy was grounded in a competitive approach to industry development, it did not mean an end to regulation. The FCC wisely determined that the nascent industry would need protection from AT&T and COMSAT if it were to have a chance of succeeding. No restrictions were placed on AT&T's ownership of satellites in support of its public telephone service. The company would be allowed to own and operate satellites for the provision of public, long distance communications. However, the FCC imposed a number of restrictions on AT&T ownership and use of satellites in the *competitive sector* of the domestic, satellite communications market.

Initially, the competitive sector of the satellite communications industry comprised: (1) TV program transmission, (2) private-line voice services, and (3) specialized services adapted for individual companies (Brock 1981, 263). In order to foster competition, AT&T was prohibited from operating in this market until one of two conditions were met: (1) until it could demonstrate that there was "substantial utilization" of any of the satellites providing specialized services, or (2) until it had been operating a satellite for three years (Bruce, Cunard, and Director 1986, 264). When either of these conditions had been met, AT&T could petition the commission for a removal of the restrictions. However, approval was not automatic. The FCC would still consider the extent of competition and whether the potential existed for AT&T to cross-subsidize satellite services. In order to prevent COMSAT from gaining an unfair advantage in the domestic market, the consortium was obliged to fully separate its domestic and international operations.

Although a number of corporations initially expressed a strong interest in establishing satellite communications systems, only a few of these projects would actually materialize. Of the eight applications the FCC received in 1970, only four were in operation by 1980. The first entrants were RCA, the electronics manufacturer, and Western Union, the common carrier. RCA launched its first satellite, *Satcom 1*, in January 1974. Western Union's entry, *Westar 1*, was launched in April of the same year. These launchings were followed shortly after by *Satcom 2* and *Westar 2*. AT&T, GTE, and COMSAT combined their resources to launch two satellites in 1976: *Comstar 1* and *Comstar 2*. The satellites were owned by COMSAT, which then leased transponders to AT&T and General Telephone. Jill Hills notes that satellite companies initially had difficulty establishing a competitive private-line voice service as a result of echo problems that plagued early satellite technology. Because of these problems, conversations could only be sent in one direction (1986, 62). The industry would have to wait until 1979 for these problems to be resolved. Also, the first satellites used C band frequencies, which required large earth receiving stations that had to be located outside of urban areas to avoid interference with microwave communications. The competitive satellite operators were dependent on AT&T for local distribution, and as a result, subject to the same kinds of delays and interconnection problems that had plagued the microwave specialized common carriers (62). More successful was the use of satellites for the one-way transmission of computerized data.

As a result of the underutilization of satellite transponders for private lines and private networks, an opening was created for the transmission of cable TV programming on the domestic satellites. During the 1970s there was a dramatic decrease in the cost of earth receiving antennae. The first antennae were thirty-five feet in diameter and cost over $100,000. By 1979 the cost had dropped as low as $10,000 for antennae with a diameter of only fifteen feet. The impact on the cable TV industry was dramatic: by 1979 there were over 2,250 receiving antennae in service throughout the United States (Brock 1981, 264). By 1979 twenty of the twenty-four transponders on RCA's *Satcom 1* satellite were being used for the transmission of cable TV programming (Tunstall 1986, 72).

The growth of the cable TV industry was reinforced by the emergence of Home Box Office (HBO), a specialty pay TV channel, and Turner Network Television (TNT), Ted Turner's Atlanta-based superstation. The first cable TV services, known at the time as Community Antenna Television (CATV), were approved by the FCC in 1955. Essentially the service involved the retransmission of over-the-air (broadcast) TV programming. Early growth took place outside urban centers where the reception of broadcast signals was poor. The basic technology consisted of an antenna, installed on a high point, such as a hill, mountain, or tall building, that was capable of receiving TV signals from the nearest city. The signals were then amplified and retransmitted by coaxial cable to the home. Early cable TV was essentially a local media, limited to a handful of stations. The introduction of microwave towers made it possible to receive TV signals from other cities, thereby increasing the number of channels available to the cable subscriber. Satellite technology accentuated this trend by making it possible to retransmit programming to a national audience.

In 1972 approximately six million homes subscribed to a cable TV service. This was less than 10 percent of U.S. homes that had TV sets (Tunstall 1986, 124). Ten years later the percentage of U.S. households with cable had almost tripled to 28 percent (Tydeman et al. 1982, 111). Although the availability of satellite transponders was an important factor, changes in FCC policy towards the cable industry also played a significant role in the growth of the industry. Historically, the FCC had attempted to protect local broadcasters by limiting the number of "foreign," that is, non-local signals a cable operator could carry on its service. This policy was designed to prevent the siphoning off of advertising revenues from small markets to large markets. Every cable operator was required to carry all local signals. The number of network stations (NBC, CBS, and ABC) and independent stations an operator could carry would depend on the size of the market. The larger the market, the more non-local stations could be carried. Gradually, throughout the 1970s, the FCC relaxed these restrictions. By 1978 most of these constraints had been removed either by the FCC or through decisions rendered in the courts (Horwitz 1989, 257).

HBO was one of the first enterprises to capitalize on the new rules and the potential of satellite transmission to reach a national audience. It did this by developing specialty programming (primarily sports and films) for sale to cable operators and their subscribers. The programming developed by HBO was not available over-the-air in any region of the country. It was produced exclusively for viewing by cable subscribers for a monthly charge. Turner's TNT was a variation on a familiar model. He gambled that there was a national audience for the sports, news, and films that were being broadcast by a failing independent TV station (WTCG) that he owned in Atlanta, Georgia. HBO began using *Satcom 1* in 1975. Turner's superstation followed in 1976, also on *Satcom 1*.

The liberalization of satellite communications followed a distinctive course, one that did not mimic or replicate the series of events that led to the liberalization of terminal equipment and long distance. In the case of satellite communication *a new technology emerged to meet demand for new services in a field that was not already occupied by AT&T*. The Nixon administration played an important role when it urged the FCC to adopt an open skies policy. Coming as it did from the executive branch, the recommendation was an unusual expression of executive volition that the FCC would have been hard-pressed to ignore. It is relevant to note that the *Domsat II* order occurred in 1972 shortly after the *Specialized Common Carrier* decision (1970), and just four years after the FCC's *Carterfone* decision (1968). A shift in policy favoring competition on the fringes of the telecommunications industry was already underway. *Domsat I* and *Domstat II* extended this policy trend into a new field.

Although AT&T and COMSAT attempted to convince NASA and the FCC that the satellite industry should be run as a monopoly, these efforts did not bear fruit. More than a decade passed between the Communications Satellite Act of 1962 and the launch of the first competitive domestic satellite in 1974 (*Satcom*

1). But the delay in establishing a policy for domestic satellites was only partially due to AT&T obstruction. Recall that ABC's request to operate a private satellite network did not come until 1965. The FCC spent the rest of the decade settling fundamental questions, such as: whether it had the authority under to the Communications Act to authorize the *private* ownership of communication satellites, how to allocate spectrum space to satellites, and what would be the impact of private ownership and a competitive industry structure on international satellite policy (Bruce, Cunard, and Director 1986, 262). Most important, in the late 1960s and early 1970s satellite technology was not sufficiently advanced to be a competitive threat to land-based transmission in the private-line, and private network markets. The first satellites were not heavily used. It was not until a mature cable TV industry began introducing specialty programming for a national market, that the glut of excess satellite capacity began to diminish.

From a regulatory point of view, the FCC's open skies policy was notable in at least two respects. First, the commission was prepared to license communication satellites without a strong presentation of evidence concerning the specific uses they would be put to and the real extent of demand for satellite services. Under the approach adopted by the commission, the market would answer these questions. This was a departure from the approach that had governed the licensing of land-based installations—one that adhered strictly to the letter of the Communications Act of 1934. Section 214 of the act stipulated that in order to receive a license to undertake the construction of any transmission line, it was necessary to obtain from the commission "a certificate that the present or future public convenience requires or will require the construction, or construction and operation of such additional or extended line." The onus was on the carrier to demonstrate that construction and operation were in the public interest, usually through the presentation of strong evidence of proven demand. This requirement was waved in the case of the first communication satellites. Second, the FCC's domestic satellite policy made a distinction between dominant and non-dominant carriers. It did this in order to protect the non-dominant carriers from the market power of dominant ones, such as AT&T. In the case of *Domsat II*, this protection took the form of an explicit prohibition on AT&T participation in the market for competitive satellite services. The *Domsat II* order was one of the first examples of *pro-competitive* regulation. Rather than protecting the monopolist from entry, pro-competitive regulation was designed to foster and protect competition by restricting the activities of dominant firms.

5.8 REGULATED MONOPOLY AND ANTITRUST LAW: A COMPLEX DYNAMIC

Whether or not regulated monopolists have immunity from antitrust is a complex question. Judicial interpretation of the Sherman Act resulted in a number of doctrines or principles that had a bearing on the activities of regulated monopolists. The following practices were judged to be anti-competitive: (1) denial of essential facilities to firms in adjacent markets, (2) tying, (3) leveraging, (4) predatory pricing and cross-subsidies, and (5) transfer pricing and self-dealing (Kellog, Thorne, and Huber 1992, 139–145). The *essential facilities* doctrine "requires a firm with monopoly power in one market to deal equitably with competing firms operating in adjacent markets that depend on it for essential inputs" (139). *Tying* is a marketing practice involving the pairing of two distinct services together. This is an appealing option for a monopolist who wishes to extend his or her monopoly into an adjacent, unregulated market (141). *Leveraging* involves the exploitation of monopoly power to gain an unfair advantage in an adjacent market. For example, a monopolist might refuse to deal with companies in an adjacent market as part of a long-term strategy designed to usurp the other market (142). *Predatory pricing* is pricing below cost in order to drive competitors out of business. It is facilitated by the use of hidden transfers (subsidies) from the regulated sector of the monopolist's business to the competitive one (143–144). *Transfer pricing* and *self-dealing* was the basis for the case against Western Electric. It entails the selling of inputs from an unregulated affiliate to a regulated parent company at inflated prices. This has the effect of distorting rate-of-return regulation (145). These practices share a common attribute: they all involve an abuse of monopoly power in an adjacent market not subject to regulation.

Under the Sherman Act, antitrust suits could be initiated by either private parties or the government—at the federal level the DOJ (by the attorney general). Following the 1956 Consent Decree a number of private antitrust suits were launched against AT&T. For example, in 1974 MCI sued AT&T for anticompetitive practices in the long distance market. Among the charges was a complaint that AT&T had failed to provide *essential facilities* when it frustrated MCI's attempts to gain access to Bell System local facilities (Kellog, Thorne, and Huber 1992, 159). Other suits involving the essential facilities principle included one filed by Southern Pacific Communications in 1978 (later to become Sprint), and one filed by Mid-Texas Communications Systems. Mid-Texas Communications sought to provide a local telephone service and required interconnection with AT&T's long distance network. MCI's suit against AT&T also included charges of *predatory pricing*. *Carterfone* began as an antitrust action. Ultimately, the case was resolved by the FCC, and not by the courts. However, once the FCC sided with *Carterfone* an "avalanche" of antitrust suits followed against AT&T related to its control of the equipment market (175). Brock notes that between the *Carterfone* decision in 1968 and 1974, no less than thirty-five private antitrust suits were filed against AT&T (1981, 295). Moreover, AT&T was not the only target of antitrust.

General Telephone and Electronics (GTE) Corporation, which provided local telephone service throughout the United States,[2] was the second largest telephone company in the United States. In 1959 GTE merged with Sylvania Electronics, an electronics manufacturer. Similar to AT&T, GTE had a vertically integrated business. After Western Electric, GTE was the second largest manufacturer of telecommunications equipment in the United States; it was also a major supplier of equipment to independent telephone companies (Kellog, Thorne, and Huber 1992, 183). International Telephone and Telegraph (ITT), the second largest manufacturer of telecommunications equipment in the world, filed an antitrust suit, *ITT v. GTE Corp.*, charging GTE with exclusive dealing between its phone companies and its manufacturing arm. In 1972 a Hawaii district court ruled in favor of ITT and ordered GTE to divest itself of its manufacturing subsidiaries. The decision was set aside on appeal. The case was tried again in 1978 with the court finding for ITT a second time. As a result, GTE was liable to ITT for damages (184). The *ITT* decision was an important encouragement to other companies seeking antitrust damages.

The record of private antitrust litigation was mixed. For example, while the MCI suit charged AT&T with twenty-two forms of misconduct, only fifteen went to jury, and only ten of these were recognized by the jury. As a result, MCI only received $600 million, two-thirds of the damages it sought (Kellog, Thorne, Huber 1992, 159). In general, the plaintiffs were most successful in cases where the FCC had already ruled, such as in the *Carterfone* decision. In cases where the FCC backed the telephone companies, the plaintiffs were not successful (181). Although private antitrust actions were annoying and costly to AT&T, they did not threaten its monopoly.

5.9 THE FINAL ANTITRUST CHALLENGE: AT&T AGREES TO DIVEST ITSELF OF ITS LOCAL OPERATING COMPANIES

In 1974 the government launched its third antitrust suit against AT&T. Recall that the 1913 challenge was settled in 1914 with the Kingsbury Commitment. The 1949 action was settled with the Consent Decree of 1956. Once again a settlement was reached before the case had run its course through the courts. However, this time AT&T agreed to divest itself of its local operating companies, something that was not even on the table in the 1956 negotiations. AT&T would disassemble its vertically integrated business. In return, it would be freed from the constraints of the Consent Decree, which had limited the company's activities to the common carriage business. The decision came as a surprise to most observers, including the over one million employees of AT&T. Not only were the terms of the settlement unanticipated, but the manner in which the case was settled was quite unorthodox for a case of this magnitude. The story begins a decade earlier when AT&T responded to the third antitrust suit by launching a campaign for new legislation that would protect its monopoly.

In the wake of *Carterfone* (1969), the *Specialized Common Carrier* decision (1971), the opening of MCI's intercity private-line service (1972), and the DOJ's latest antitrust suit, AT&T began to prepare a legislative defense of its monopoly. The company rallied its unions, the independent telephone companies, and state regulators in support of legislation that would have reaffirmed the intent of the Communications Act of 1934 that competition in telecommunications was not in the public interest. The bill was known as the Consumer Communications Reform Act of 1976. AT&T launched a major public relations campaign and lined up 175 members of the House of Representatives and 16 senators in support of the bill (Brock 1994, 151). But the bill did not have the support of Congressman Lionel Van Deerlin, who was the chairman of the House Subcommittee on Communications. Van Deerlin was not prepared to allow the bill to move smoothly through his subcommittee. Instead, he proposed extensive hearings on the subject and in 1978 produced his own bill that was pro-competitive. The tabling of the Van Deerlin Bill put an end to any hopes AT&T might have had that monopoly legislation would pass easily through the U.S. Congress. A number of pro-competitive bills were introduced between 1978 and 1980 without success. For all intents and purposes the legislative approach had resulted in a stalemate between the proponents of competition and proponents of monopoly.

On June 22, 1978, the DOJ's antitrust case was reassigned to Judge Harold Greene of the Federal District Court of Washington, D.C. During the previous four years the case had moved forward at a snail's pace. Among the factors that contributed to the slow advance of the pre-trial proceedings was the fact the judge assigned to the case had become terminally ill. Greene, who was observing his first day as a judge on the federal bench, was eager to demonstrate that he was capable of moving the case expeditiously through the court. He quickly imposed an accelerated calendar for the pre-trial proceedings. Moreover, Greene had little tolerance for AT&T's procedural delays. As the 1970s drew to a close, the DOJ was prepared to offer a settlement to AT&T "on easy terms requiring only a token divestiture" (Derthick and Quirk 1985, 200). The offer was prompted by the high turnover of DOJ attorneys and the department's inability to match the considerable resources that AT&T had assigned to the case. In what turned out to be an important miscalculation, AT&T rejected the offer because it falsely believed that the pro-business Ronald Reagan administration, which would be taking office in January 1981, would reign in the "zealot" at the Justice Department.

The conservative Reagan administration was elected on a campaign promise to reduce government interference with business. Unfortunately for AT&T, this pledge did not extend to the DOJ under the stewardship of William Baxter, the new attorney general. Before joining the Nixon cabinet, Baxter had been a professor of antitrust law and was knowledgeable of economics. Baxter was determined to press the case against AT&T despite opposition from a number of his colleagues in the Reagan cabinet. In a familiar pattern the Department of Defense strongly opposed a continuation of the DOJ's suit. Also aligned against antitrust was the Department of Commerce, which feared that deregulation would open the U.S. market to foreign terminal equipment without a reciprocal opening of foreign markets to U.S. products (Brock 1994, 159).

The stalemate between Baxter and his cabinet colleagues produced two results: (1) the DOJ continued to pursue its case to trial before Judge Greene, and (2) a final attempt at a legislative solution was proposed by Republican senator Bob Packwood. Had it passed, the legislation would have separated AT&T's basic, regulated services from its competitive services. Accounting rules would have been introduced to ensure that there were no transfers between regulated and unregulated businesses. Rules would have been introduced to ensure that AT&T would share network information with its competitors and not favor its own affiliates (Brock 1994, 160). The FCC would have been assigned an important role in ensuring that network information circulated freely between AT&T and its competitors. Supporters of the bill included AT&T, the FCC, and the Defense and Commerce Departments. Opposed to the bill was MCI. Moreover, the DOJ was concerned that it did not go far enough to control AT&T's market power (160). In October 1981 the bill passed the U.S. Senate with nearly unanimous support. But the bill failed to pass in the House of Representatives.

When the Packwood Bill failed to pass through Congress, attention was again focused on the court arena. The trial phase of the case had begun on January 15, 1981. It took the DOJ four months to present its case before Judge Greene, calling over one hundred witnesses and presenting thousands of documents (Kellog, Thorne, and Huber 1992, 208). After the government had presented its case, AT&T, in a strategic move, appealed to the court for a complete dismissal of the case. On September 11, 1981, Judge Greene responded to AT&T's motion to dismiss in an unusually lengthy opinion, which made it clear that he found merit in the government's case that AT&T had unlawfully attempted to exclude competition from long distance. AT&T had not even presented its case (it had simply moved for a dismissal), and it was already clear that the DOJ would receive favorable treatment from the judge. The handwriting was on the wall. This, combined with the failure of the Packwood Bill, made it clear to AT&T that it was in trouble. AT&T's chairman, Charles Brown, determined that it was time to cut a deal and reduce its losses. This time AT&T would not be able to avoid a breakup of its vertical monopoly. It was time to advance a negotiated settlement.

In the early 1980s there was a broad consensus that local networks were a natural monopoly. Attorney General Baxter's plan was relatively simple: he wanted an agreement based on the full separation of AT&T's monopoly business from its competitive business. This would be accomplished by requiring AT&T to divest itself of its local telephone companies. Baxter had claimed that he could settle the dispute with a two-page decree. Essentially, this is what happened: On December 21, 1981, the DOJ presented a first draft of a settlement. Negotiations ensued, which resulted in an announcement on January 8, 1982, that the government and AT&T had reach an agreement. The trial would not proceed. The settlement was referred to as the Modification of Final Judgement (MFJ) due to the fact that both AT&T and the DOJ considered it to be a modification of the 1956 "Final Judgement," the Consent Decree. Under the terms of the agreement, AT&T's assets would be divided in two: long distance service, Western Electric, and Bell Laboratories would remain under the AT&T umbrella; and the local operating companies would be separated from AT&T. In return for relinquishing control of its local telephone monopoly, AT&T would be released from the constraints of the 1956 decree, which had prohibited the company from participating in non-common carrier businesses such as computing and information services.

What happened next could not have been anticipated by either Attorney General Baxter or AT&T chairman Brown. Although the 1956 settlement had been entered in a New Jersey district court, the 1974 antitrust case had gone to trial in the Federal District Court of Washington, D.C. The old case was in a New Jersey court; the new case was in a Washington, D.C., court. Moreover, while the 1974 antitrust case was going to trial, a separate appeal of the 1956 decree was underway in New Jersey District Court. AT&T had launched the appeal to determine whether the Bell companies could, under the 1956 decree, offer information services. In order for the new settlement to have legal standing, a motion to withdraw the 1974 case had to be filed by the parties in Judge Greene's court in Washington. However, in order for it to be considered as a "modification" of the 1956 "final judgement," the New Jersey court, which had jurisdiction over the 1956 settlement, had to agree to transfer the original case to Judge Greene's court. Moreover, AT&T, whose appeal of the 1956 settlement was pending in New Jersey District Court, had to file a motion before the New Jersey judge signaling that they were withdrawing their appeal.

When the DOJ and AT&T presented the new settlement to the New Jersey judge, Vincent Biunno, he simply signed it, based on his assumption that the matter fell under his jurisdiction. It was only a week later that Biunno was persuaded that his signature was required to transfer both the 1949 case (which resulted in the 1956 decree) and the new decree to Judge Greene's court (Kellog, Thorne, and Huber 1992, 212). He did so reluctantly. However, when the parties attempted to file the new decree in Washington, D.C., they discovered that Greene was not prepared to enter it without review. The MFJ was an agreement between two parties: AT&T and the DOJ. Therefore, it did not require approval by Congress or the FCC. It was an altogether different matter whether the settlement required the approval of the court. Judge Greene decided that it did, basing his decision on an interpretation of the Tunney Act (1974). Also known as the Antitrust Procedures and Penalties Act, it stipulated that consent decrees reached in antitrust cases were subject to review by the court to determine if they were in the public interest. AT&T and the DOJ argued that the settlement was not subject to judicial review because *it was simply a modification of a previous agreement*. Although clever, and no doubt convenient, the fine points of the argument were lost on Judge Greene, who interpreted the situation differently and proceeded to place his stamp on the AT&T divestiture.

The initial settlement agreement between the DOJ and AT&T would have reorganized AT&T in the following manner:

1. AT&T would divest itself of its twenty-two Bell Operating Companies (BOCs). These companies would then be consolidated into seven Regional Bell Operating Companies (RBOCs) (fig. 5.1). For example, New Jersey Bell, Bell Telephone of Pennsylvania, Chesapeake & Potomac Telephone, C&P of Maryland, C&P of Virginia, and C&P of West Virginia were combined to form the RBOC, Bell Atlantic[3]
2. The BOCs would only be allowed to offer local telephone service. They would not be allowed to manufacture or sell telephone equipment, provide a long distance telephone service, provide an electronic information service, or Yellow Pages
3. A Central Services Organization would be created to coordinate the distribution of technical information between the BOCs
4. AT&T would retain control of AT&T Long Distance, Western Electric, and Bell Laboratories (its research subsidiary)
5. AT&T would assume ownership of Yellow Pages and all equipment leased to Bell System customers
6. The BOCs would be required to provide equal access to their facilities to all long distance carriers; no preferential treatment would be given to AT&T
7. AT&T would no longer be restricted to the business of common carriage as prescribed in the 1956 Consent Decree, as such, it would be free to enter computer markets and information services (Simon 1985, 26–27)

After hearing testimony from interested parties, Judge Greene determined that the settlement went too far in limiting the scope of business of the newly created RBOCs. His principal concern was that they would be confined to the most capital-intensive part of the business with few prospects for growth. If the BOCs were not sustainable as business entities this would have a negative effect on the quality and affordability of local service. Accordingly, he revised the agreement to expand the scope of the RBOCs business to include the operation of Yellow Pages and the sale of equipment (telephones, switches, and so on) to customers (fig. 5.2). However, the RBOCs were not allowed to manufacture equipment. Conversely, with respect to AT&T, Greene moved to temporarily restrict the company's entry into electronic publishing, where AT&T could exercise control over the content of messages transmitted over its network. The restriction on electronic publishing was broad, encompassing radio, TV, and interactive transactional services. AT&T would have to wait seven years (until 1989) before it could enter these businesses.

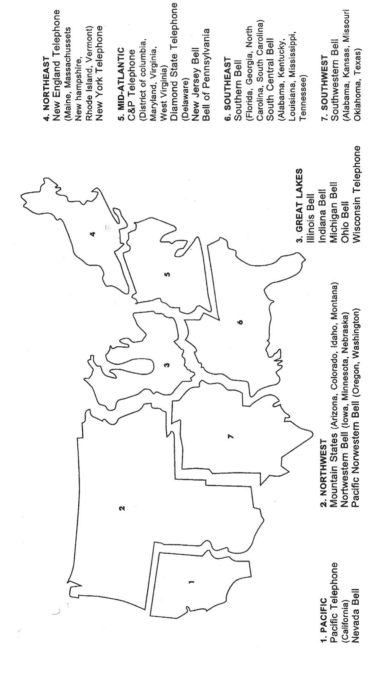

4. NORTHEAST
New England Telephone
(Maine, Massachussets
New hampshire,
Rhode Island, Vermont)
New York Telephone

5. MID-ATLANTIC
C&P Telephone
(District of columbia,
Maryland, Virginia,
West Virginia)
Diamond State Telephone
(Delaware)
New Jersey Bell
Bell of Pennsylvania

6. SOUTHEAST
Southern Bell
(Florida, Georgia, North
Carolina, South Carolina)
South Central Bell
(Alabama, Kentucky,
Louisiana, Mississippi,
Tennessee)

7. SOUTHWEST
Southwestern Bell
(Alabama, Kansas, Missouri
Oklahoma, Texas)

3. GREAT LAKES
Illinois Bell
Indiana Bell
Michigan Bell
Ohio Bell
Wisconsin Telephone

2. NORTHWEST
Mountain States (Arizona, Colorado, Idaho, Montana)
Nortwestern Bell (Iowa, Minnesota, Nebraska)
Pacific Norwestern Bell (Oregon, Washington)

1. PACIFIC
Pacific Telephone
(California)
Nevada Bell

Figure 5.1   Reorganization of the Bell System Companies into Seven Regional Holding Companies

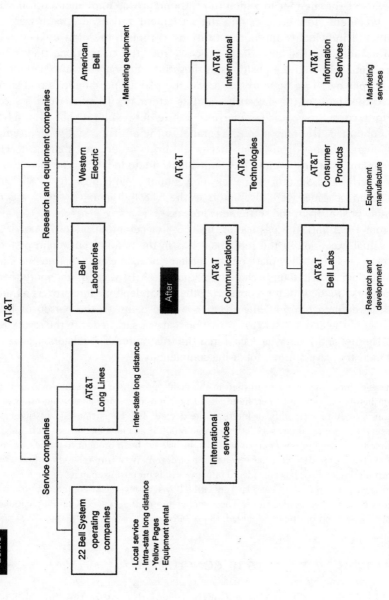

Figure 5.2 AT&T before and after the Divestiture

The AT&T divestiture was a unique chapter in the history of U.S. telecommunications, both in terms of the process leading up to the settlement, and in terms of the manner in which the settlement was implemented following the agreement. To a great extent, the settlement and its aftermath were the product of individuals and not institutions. Although the individuals derived their standing in the divestiture process from the institutions they represented and reflected the positions of these institutions, they also had discretionary powers. To a large extent it was the play of personalities that determined the shape of the divestiture and its aftermath. In other words, were it not for Baxter, Brown and Greene, a very different result probably would have been obtained. Running a course parallel to the antitrust suit were attempts by Congress to pass legislation that would bring order and direction to U.S. telecommunications policy. Key individuals included Congressman Lionel Van Deerlin and Senator Bob Packwood. Finally, William Baxter's colleagues in Reagan's cabinet were also actors in the AT&T drama. The FCC was noticeably absent from the settlement process.

From 1982 until the passage of the U.S. Telecommunications Act of 1996, the actual execution of the settlement was in the hands (courtroom) of Judge Greene. Greene put his mark on the settlement to an extent that no one could have anticipated. It soon became clear that the MFJ was nothing more than a framework, a blueprint for reorganizing the U.S. telephone industry. As a framework it required *constant interpretation* by the court, in the person of Judge Greene. In the end it is safe to say that no one was satisfied with the result. Kellog, Thorne, and Huber have captured the frustration of regulators, legislators, and industry players in the following commentary:

> Indeed, more paper has been filed in the eight years since divestiture than in the ten before. There have been hundreds of new motions, complaints, and other requests to enforce, modify, or interpret the decree. The Department of Justice has issued thousands of advisory letters. Some 6,000 briefs have been filed with Judge Greene. Fifteen groups of consolidated appeals have been brought to the D.C. Circuit. The Supreme Court has received a half dozen divestiture-related petitions for certiorari. Congress has considered numerous bills proposing changes to the decree. At least a dozen times a year, lawyers and lobbyists converge at conferences to discuss pending decree decisions. The membership of the Federal Communications Bar Association has nearly doubled. (344–345)

5.10 CONCLUSION

As AT&T entered the second half of the twentieth century, the three pillars of its station-to-station, network monopoly were: (1) the manufacture and provision of terminal equipment; (2) the provision of long distance service; and (3) the provision of local access and local exchange service through its subsidiary operating companies. In this chapter we have related the story of the liberalization of the terminal equipment and long distance markets and the divestiture of local access and exchange service from AT&T's empire. Today, AT&T supplies telecommunications equipment in a competitive market and provides long distance services in a competitive market. But the company is no longer in the business of providing local network services. The markets for telecommunications equipment and long distance telephony are subject to competition. The liberalization of the local exchange market is pending.

How did the United States go from monopoly to competition in two of the three pillars of the telecommunications industry? Was the transformation from monopoly to competition preordained by the advance of new technologies? If AT&T had managed events differently, more effectively, could it have achieved a different result? Could the industry have followed a different path towards liberalization, and if it did, would it have produced a different result?

In retrospect, it would appear that the introduction of competition to the equipment and long distance markets was inevitable. Technology clearly played a role in creating opportunities for entry into niche markets, but so did business demand for new services and for lower rates. The two were interrelated. On the negative side, AT&T's inability to respond in a timely fashion to the demands of America's largest corporations frustrated an important clientele, and lent credence to calls for liberalization. AT&T's inertia may have been a function of a proud, but also arrogant, "monopoly-oriented," corporate culture—an organizational culture cultivated during more than a half a century of operations. Indeed, this corporate culture, based on the slogan "One system, one policy, universal service," served the company well. AT&T's corporate culture clearly had problems adapting to the changing business environment defined by emerging computer, data processing, microwave, and satellite technologies. On the other hand, AT&T's monopoly (in its dual aspect as both corporate form and corporate philosophy) enabled the company to stave off competition for over thirty years. No mean accomplishment.

There was no fatal blow to AT&T's monopoly. It was not necessary. Under pressure from microwave, satellite, and computer technologies, the veils were slowly, incrementally, torn from the sham argument of natural monopoly. At one time, AT&T's vertically integrated business may have been an efficient model for building and operating a network. This was particularly true during the early phase of network growth when local access to exchange facilities was virtually nonexistent. These local connections (local access) had to be built from scratch, requiring a significant capital investment. But no one has ever been able to demonstrate definitively that the relationship between the three pillars of AT&T's monopoly was characterized by economies of scale and scope. This is not to say that the vertically integrated nature of AT&T's business did not serve social policy objectives. The vertical integration of services, when combined with value-of-service pricing and system-wide price averaging facilitated revenue transfers between the different classes of services as part of a policy to promote universality. Although vertical integration may have facilitated these transfers, it was not a necessary condition for their effectiveness.

To summarize, there was nothing "natural" about AT&T's vertically integrated monopoly. The historical record suggests that it was the "unnatural" result of an informal alliance between regulator and regulated. Although unnatural (in the economic sense), this alliance was comprehensible, and even predictable, given the fuzzy nature of the FCC's legislative mandate. As Kellog, Thorne, and Huber have suggested, the Communications Act of 1934 placed a heavy burden of proof on new entrants to demonstrate that the construction of new facilities was in the public interest. While the act gave the FCC the power to sanction competition by authorizing the construction of facilities and ordering interconnection, it did not endorse or otherwise encourage a competitive industry structure. In the absence of changes to its legislative mandate, or other forms of guidance from elected authority, the commission merely adopted a prudent approach to change, based on a conservative interpretation of its mandate. The FCC may have been slow to change, but it was not intransigent. Even before the Telecommunications Act of 1996, the FCC had become an ardent proponent of competition, thus demonstrating that the agency was capable of adapting itself to changing technology, changing industry conditions, and a changing political climate associated with the movement to reduce government interference in the economy.

Did the AT&T divestiture spell the end of monopoly and the end of regulation? Far from it. Monopoly is still present at the level of local access and local exchange service. The divestiture, while disassembling AT&T's *vertically integrated* monopoly, left intact the local service monopolies, albeit in the hands of the former subsidiary BOCs. The final chapter in the movement to liberalize telecommunications markets is in the process of being written. It concerns the liberalization of local access and local exchange service. Proponents of competition take encouragement from the successful liberalization of terminal equipment and long distance markets.

As we will see in subsequent chapters, the liberalization of the local service market is a current preoccupation of policy makers in the United States and Canada. It remains to be determined, however, whether the experience of liberalized terminal equipment and long distance markets is applicable to the market for local services. The presumption underscoring current telecommunications policy, both in the United States and Canada, is that it is. In other words, current policy assumes that competition in the provision of local access and local exchange facilities is both economic and sustainable. It is our contention that the success of earlier liberalization may not necessarily apply to the local service market. Rather, the easy phase of liberalization is over. The next phase will be more difficult. In this case, the past may not serve as a prologue to the future. As such, policy makers may have to reconsider their assumptions about the sustainability of competition and the non-existence of monopoly conditions in local markets if they are to design effective policy. To the extent that past experience in the liberalization of telecommunications is an important factor shaping current telecommunications policy, it is important for the student of communications to be familiar with the story of early liberalization.

NOTES

1. "According to the agreement, Intelsat owns and operates the space segment of the network that includes satellites, tracking, and telemetry control facilities as well as the administrative machinery coordinating procurement, planning, and financial management. Revenues are derived mainly from utilization charges and, 'after deduction of operating costs, are distributed to signatories—in proportion to their investment shares—as repayment of and compensation for use of capital.' Thus each member country owns a portion of the space segment equivalent to its investment share and traffic volume, and it also owns the earth segment—antennas, receivers, transmitters, and other ancillary equipment. Because Intelsat was founded on the premise that access to the network would be available to all nations on an equal basis, members and nonmembers alike are charged the same rates for the same service. Tariffs actually charged to customers, however, are determined solely by the member countries. Therefore, while Intelsat's charges to a country remain the same, what the country charges the end user may vary" (Akwule 1992, 59).

2. GTE was also active in Canada. In 1960 it acquired a controlling interest in British Columbia Telephone Company. In 1966 it gained a controlling interest in Québec-Téléphone.

3. The other six RBOCs are NYNEX, Bell South, Southwestern Bell, Ameritech, US West, and Pacific Telesis.

6

Local Network Competition and the Deregulation of Enhanced Services in the United States

6.1 INTRODUCTION

In chapter five we examined the series of events that lead to the breakup of AT&T's vertically integrated monopoly. The liberalization of the terminal equipment and long distance markets, when combined with the divestiture of AT&T's local operating companies, put an end to AT&T's station-to-station monopoly. But it left intact the Bell System local operating companies' stranglehold on the local exchange market. The terms of the Modification of Final Judgement (MFJ) ensured that the Bell Operating Companies (BOCs) could not reconstitute a vertically integrated business: they were not allowed to provide long distance service or to manufacture terminal equipment for sale or lease to their subscribers. In this chapter we will recount how the last vestiges of telephone monopoly, the market for local exchange service, was opened to competition. We will see how the Federal Communications Commission (FCC) accomplished this objective by linking a policy for competition in basic telephone service to a resolution of the problem of telephone company participation in the provision of enhanced services, that is, services based on applications which combined computing and telecommunications.

6.2 THE MARRIAGE OF COMPUTERS AND TELECOMMUNICATIONS

The first computers designed for commercial (as opposed to scientific) use began to appear on the market in the mid-1950s. For the most part, these computers were used to automate the routine processing of information in large organizations: preparing payrolls and checks, processing bills in telephone and utility companies, and recording and sorting bank checks. Eventually computer-based data processing would be used to record orders from customers and to manage a company's inventory.

Almost immediately, applications were developed that combined data processing with communications. For example, as early as 1956 Sylvania Electronics Products used telecommunications lines to link terminals in fifty-one American cities for data entry with a centralized data processing center in Camillus, New York ("The Computer Age" 1957, 59). One of the best known computer-communication applications was an automated reservation service developed by International Business Machines (IBM) for American Airlines. Known as the Semi-Automatic Business Related Environment (SABRE), the system connected twelve hundred remote terminals to American Airline's central database using twenty thousand kilometers of dedicated telecommunications lines (Birrien 1990, 61).

The system developed for American Airlines was a typical "time-sharing" application of the early period of computer development. The third generation mainframe computers were large, expensive, and centrally located machines. These computers were fast enough to simultaneously run several applications for several different users. Although the computer was executing commands (from different applications) sequentially, it was so fast that the individual user was unaware that he or she was "sharing time" on the computer with others. Initially, users worked from terminals that were located in proximity to the mainframe computer. But there was no technical reason for the terminal to be co-located with the computer. The computer could be controlled from dumb terminals in remote locations using leased-lines provided by a telephone company. In many cases, the only way to justify the high cost of the mainframe computer was to maximize its use through sharing, either with other divisions of the same company, or by leasing time on the computer to other companies. The marriage of telecommunications and computing was often the most economical way to harness the power of the third generation, mainframe computers.

Computer data processing (including remote data processing using telecommunications) was developed to automate the processing of information that occurs routinely in all organizations. These systems were developed in support of a company's primary business, which typically was not communications or data processing. These support activities could be operated within a company, or provided by an outside firm specializing in computer services. The early users of these computer applications were companies, not individual consumers.

With the introduction of the personal computer in the late 1970s, the information processing power of the computer was made available to the individual consumer. The computer was no longer simply a support technology for business; it became a mass-market commodity. In a similar fashion, the information processing power of the computer has been the impetus for a number of innovative communication services that are currently offered either directly by the telephone companies, or by other firms, in conjunction with local access service (local dial-up connections) provided by the telephone companies. At one end of the spectrum are the enhanced call management services offered by the telephone companies. These voice services include: call answering, caller identification, call forwarding, multi-point conferencing, and call blocking, to name a few. At the other end of the spectrum are digital communication services such as electronic mail (e-mail) and access to the Internet. These enhanced communication services, whether voice or data, rely upon the information processing power of the computer.

When computers were married to telecommunications as the basis for new services to organizations, regulators were faced with a singular problem: How would they draw a line separating the business of communications common carriage (which is regulated) from the business of information processing (which is unregulated)? The first issue is why draw a line at all? The response is unfair competition. More specifically, it is referred to in antitrust law as the denial of essential facilities (*see* chapter five, section eight). When a common carrier provides a remote data processing service in a competitive market, it provides inputs (in the form of basic transmission and switching) to both itself and its competitors. It is reasonable to assume that the common carrier will attempt to stifle competition by providing its competitors in the competitive data services market with a less reliable, or more expensive, basic communications service. If the common carrier was not a monopolist, there would not have been a problem, since the providers of the competitive services could look elsewhere for their communication inputs.

When faced with this situation, the initial regulatory response was to draw a line between computing and communications, between information processing and basic communications. In theory, once the line had been drawn, it would then be possible to tell the common carriers what types of businesses they could legally engage in. Drawing the line had an additional advantage: depending on where the line was drawn, it would enable the regulator *not to regulate* the computer/data processing industry. However, if the regulator designated information processing activities that were essential to the provision of basic communications service as part of a regulated activity, then companies offering computer services would find themselves regulated alongside common carriers when they provided equivalent services.

Computer or communications? Data processing versus communications. At one level this would appear to be a bogus problematic. Surely, it is possible to differentiate the processing of payroll information by computer from basic telephony. Data processing typically involves "the gathering, filing, storing, manipulating or using of information" (Berman 1977, 149). Establishing an electrical circuit between two callers using copper wires and a switch is the basis for telephony. What could be more different? However, consider for a moment what is occurring in the switch. The switch is actually an information-processing machine.

In the first telephone systems, calls were connected manually by a telephone operator. In these systems the processing of information was performed by a human being. Second generation switching technology was electromechanical. The building blocks of these systems were electromagnets and electromagnetic

relays. Two kinds of electromechanical switches were in use before the invention of the computer: strowger switches (invented by Almond Strowger of
Kansas City in 1889) and crossbar switches (invented by L. M. Ericsson of Swe-
den). With the invention of the electromechanical switch, it was no longer necessary for human beings to process signaling information.[1] This was done by machine. In the mid-1960s, shortly after the invention of the transistor, computers
began to be used to control crossbar switches. In other words, the switching
mechanism was still electromechanic (the crossbar) but it was controlled by mi-
croprocessors. These control systems consisted of "computer processors executing machine instructions based upon software commands stored in electronic
memories" (National Telecommunications and Information Administration
1991, 110). Because the instructions could be changed, these were referred to
as "programmable" or "stored-program" switches. The first generation of stored-
program switches were analog, and the second generation, which began to appear in 1976, were fully digital (111).

Let us return to the example of the call management services currently available from the local telephone companies. When a telephone company computer
records a voice message, displays a caller's name on, or automatically reroutes a
call to a third number, it is processing information. It is the data processing capabilities of the new generation of fully digital switches that enable the common
carrier to provide these "enhanced" voice services. When the precursors to these
enhanced services began to emerge in the 1960s and 1970s, regulators were generally unprepared to deal with them. The existing regulatory framework prescribed that the telephone companies were in the business of common carriage,
not data processing. Recall that the 1956 Consent Decree prohibited AT&T from
operating in unregulated markets. Data processing was an unregulated market.

We have just described the problem of regulatory barriers to common carrier
participation in "hybrid" services that combined data processing with communications. At the other end of the spectrum were computer manufacturers and
computer service companies, such as IBM, who wanted to offer services entail-
ing a communications component—one that substituted for functions provided
by the telephone companies. When they tried to bring these services to market
they often found that AT&T would not cooperate. When this occurred they
turned to the FCC for relief. We will now consider the complex and tangled
story of regulatory efforts to manage the boundaries between two technologies
and two industries: telecommunications and computing.

6.3 THE REGULATORY RESPONSE TO HYBRID
APPLICATIONS: *COMPUTER I*

One of the first clashes between a computer-based service provider and the regulated common carriers involved the Bunker Ramo Corporation. In the early

1960s, Bunker Ramo provided a stock quotation service to stockbrokers throughout the United States. The service involved the processing (updating) of information on stock prices and the transmission of this information to its clients (the stock brokers) using leased-lines provided by AT&T and Western Union. When Bunker Ramo attempted to expand its service so that its clients could communicate with one another in order to sell and purchase stocks, AT&T and Western Union withdrew their leased-lines (Berman 1977, 153). From Bunker Ramo's point of view, the new service was a natural extension of the service it was already providing to its clients. From the point of view of the common carriers, however, Bunker Ramo was using leased-lines to provide a switched communication service—also known in the language of telephony as *message switching*. Providing switched connections between the users of a communications network was the business of the telecommunications common carriers. It was part of their monopoly.

In 1965 the FCC began an inquiry into the relationship between the computer services industry and the communications industry. The first matter before the commission was whether, and under what conditions, the common carriers should be allowed to provide data processing services. The commission determined that common carriers could provide data processing through subsidiaries that were *structurally separate* from the parent company. Among other things this meant that the subsidiary's accounts had to be totally separate from the parent company. The subsidiary could not be represented by the parent company, nor could it use the parent company's name or corporate symbols to market services. Finally, the parent company could not purchase data processing services from the subsidiary (Berman 1977, 156). This framework only applied to common carriers such as GTE and Western Union. It did not apply to AT&T and its subsidiaries because the Bell System companies were prohibited from providing data processing, or any other *unregulated* service, under the terms of the 1956 Consent Decree. GTE's data processing subsidiary challenged the framework in the courts and won. This forced the FCC to loosen its requirement of maximum structural separation of parent and subsidiary. For example, as a result of the court ruling, the GTE parent company was allowed to purchase data processing services from its subsidiary.

The separate subsidiary rules only dealt with half of the problem. Specifically, they did not address the fundamental question of what was data processing (unregulated) and what was message switching (regulated). In other words, they did not address the question of where, and how to draw the line between the two. This was at the heart of the Bunker Ramo case. Here the issue was not common carrier participation in data processing, but whether a data processor (Bunker Ramo) was engaging in a regulated activity, common carriage.

The commission's solution, known as the *Computer I* decision (FCC 1971), was based on the premise that one could measure and compare the relative amounts of data processing and communications in a hybrid service.

The distinction between data processing and communications was a *functional one*. Although rooted in the technologies employed, it emphasized the functions they served. A simplified analogy would be the case of water delivered through a faucet to the kitchen sink. One can drink the water or it can be used to wash hands or the dishes. When we drink water its function is nourishment; when we wash with it, its function is cleansing. In our analogous example, the functions are clearly distinct. Unfortunately for the FCC, the functional distinction between data processing and communications did not hold up as well as our analogous example of nourishing and cleansing. Between pure data processing and pure communications there was a vast grey zone of activities that comprised both functions.

The *Computer I* decision established four broad categories of services: (1) pure communications (i.e., message switching); (2) hybrid communications; (3) hybrid data processing; and (4) pure data processing. Pure communications and hybrid communications would continue to be regulated. Hybrid data processing and pure data processing would remain unregulated. Hybrid communications

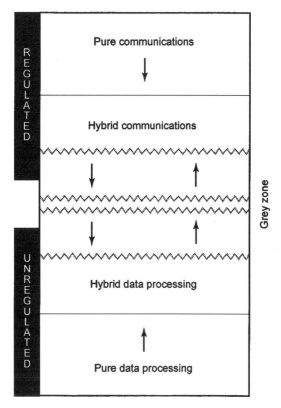

Figure 6.1   The FCC's *Computer I* Framework

and hybrid data processing constituted the grey zone of services that defied easy categorization (fig. 6.1). According to the FCC's definition, "hybrid communications" were services that were *mostly message switching* with some incidental computing. Conversely, "hybrid data processing" was *mostly data processing* with some incidental message switching (Kellog, Thorne, and Huber 1992, 543).

Whether or not a hybrid service should be regulated depended on one's ability to determine whether the principal function was data processing or message switching, and whether the service was mostly data processing or mostly message switching. By definition a hybrid service combined both functions and there was no objective test to determine which function was predominant. The FCC complicated matters by declining to give any illustrative examples of how it would draw the line. Instead, the commission proposed to regulate (or not to regulate) on the basis of ad hoc evaluations of the different services (Kellog, Thorne, and Huber 1992, 543). It is important to note that in the case of the Bell System companies, a decision *not to regulate* a service (the case of hybrid data processing and pure data processing) meant that the companies would be prohibited from offering the service. With respect to the other common carriers, which were not bound by the terms of the Consent Decree, they could only provide the service through a separate subsidiary. This framework produced some rather bizarre results with respect to AT&T. For example, if as a matter of policy the FCC were to determine that it was in the public interest for the Bell System companies to provide a hybrid service, this could only be achieved by extending the scope of the commission's authority to regulate. It is, no doubt, one of the small ironies of this period of regulation, that the FCC would have had to expand the scope of regulation if it wanted to promote AT&T participation in the enhanced services industry. In effect, common carrier participation could only be achieved at the expense of an expanded role for the FCC. By the same token, if the FCC determined that a service was hybrid communications, the competitive, computer service provider might find itself regulated as a common carrier.

6.4 THE FAILURE OF *COMPUTER I* RESULTS IN A NEW APPROACH: *COMPUTER II*

Although the FCC's intention was to produce a framework that would introduce stability to markets for the new class of computer-communication services, its efforts produced the opposite result: confusion. Moreover, it did little to promote the growth of new services. On the one hand, it had the effect of freezing common carriers out of new markets where economies of scale and scope may have existed between enhanced and basic communications services. On the other hand, the framework had a chilling effect on enhanced providers, who were not common carriers, and did not wish to be caught in the web of regulation as the result of a regulatory decision that a particular service involved more communication than data processing. It took seven years for the *Computer I* decision to be contested in the courts. For the most part the FCC's decision was affirmed by the courts on appeal. But this resolved little, since in the interim, a new generation of computer technology emerged that made the four categories even more difficult to apply, if not totally obsolete.

In 1976 the FCC initiated another computer inquiry. It lasted four years, producing the FCC's second computer decision in 1980. This decision was known as *Computer II* (FCC 1980). The new rules were prompted by a fundamental shift in computer technology from centralized to decentralized computing systems. The shift was caused by the miniaturization of computer components: from transistors to integrated circuits, and eventually to microchips. Miniaturization reduced the costs of computing while at the same time increasing the processing speed of the computer. In the age of the large mainframe computer, terminals were "dumb": they had no capacity to process information. Miniaturization made it economical to decentralize computing by locating some functions in terminal equipment. This produced "intelligent" terminals, and other intelligent peripheral equipment, and gave rise to distributed computer systems. In a distributed system computing tasks are shared between a number of computers that are connected via a network.

Shortly after the FCC issued the *Computer I* decision in 1971, AT&T decided to challenge the rules by introducing a new piece of terminal equipment known as Dataspeed 40/4. The Dataspeed 40/4 was an intelligent terminal capable of performing some information processing while communicating with a mainframe computer (Brock 1994, 94). When AT&T filed a tariff with the FCC for the new terminal, IBM objected that the Dataspeed 40/4 was more computer than communications. In other words, according to IBM, AT&T had crossed the line that separated hybrid communications from hybrid data processing. Either AT&T was operating in unregulated markets outside its mandate as a common carrier, or the company was extending its regulated activities into the previously unregulated data processing market.

At first the FCC sided with IBM. However, upon review it allowed the Dataspeed 40/4 tariff to stand pending a reconsideration of its rules on customer premises equipment (CPE). Both the Dataspeed 40/4 terminal and the basic telephone apparatus are examples of customer premises equipment. The reader will recall that by 1978 AT&T had exhausted all its attempts at preventing subscribers from owning their own telephone sets. By 1980 the FCC was prepared to liberalize the entire CPE market, including the data terminal market. Under the new competitive framework, AT&T would be allowed to provide CPE through a "fully separate subsidiary" (Brock 1994, 96). The decision to fully liberalize CPE was one component of the *Computer II* decision. The other component was a decision to liberalize *all enhanced services*. No longer would the commission attempt to make fine-grained distinctions between hybrid communications and hybrid data processing. However, the commission was not able to depart from the practice of drawing lines between different categories of services.

In its *Computer II* decision, the FCC created a boundary between regulated and unregulated activities based on the concept of *basic transmission service*. Basic transmission service would continue to be regulated. All services that *enhanced* basic service, by adding value to it, would be offered in a competitive market. CPE would be offered competitively. The *Computer II* decision was designed to radically curtail the scope of regulation. In effect, it was designed to remove the FCC from the business of regulating *anything but basic communication service*. Basic service was defined as "a pure transmission capability over a common communications path that is virtually transparent in terms of its interaction with customer supplied information" (FCC 1980, at 420). In other words, in a basic communications service the message, whether voice or data, would not be subject to any changes. In a basic service, messages would not be processed or otherwise transformed or manipulated during transmission.

Under the *Computer II* rules, everything that was not basic transmission was considered an enhanced serviced. Enhanced services would be unregulated. In order to avoid confusion, the commission provided a definition of the types of activities that would constitute an enhanced service. According to the commission, an enhanced service was one that "acted on the format, content, code, protocol or similar aspects of the subscriber's transmitted information, or provided the subscriber additional, different, or restructured information, or involved subscriber interaction with stored information" (FCC 1980, at 387). The following were examples of basic services under *Computer II*: ordinary telephone service, private lines for voice, video, and data; and the resale of basic service. Enhanced services included: business management systems for inventory; text editing services; information retrieval services (electronic mail and databases); and data processing (Bolter, McConnaughey, and Kelsey 1990, 90).

Under the *Computer II* rules the distinction between basic and enhanced service applied to all common carriers. All basic communication services provided by common carriers would continue to be regulated as prescribed by Title II of the Communications Act (1934). (Title II was the subchapter of the U.S. Communications Act that dealt with the regulation of common carriers.) All common carriers would now be allowed to offer enhanced services. But the FCC created a special framework for the provision of enhanced services by AT&T.

Because of its status as a dominant carrier, AT&T was singled out by the commission for special regulatory treatment. AT&T would be subject to a maximum separation of its basic and enhanced services similar to the maximum separation requirement set out in *Computer I*. In specific terms, "maximum separation" meant that the separate subsidiary had to own and locate all its switching and computer equipment on its own premises. It could not use equipment owned by the parent telephone company or located on the parent's premises. Separate books of accounts were to be kept by the parent and the subsidiary. The parent and the subsidiary could not share officers or any other personnel used in the installation, operation, or maintenance of services. Marketing and product advertising had to be done separately. Moreover, the subsidiary was prohibited from owning its own transmission facilities. These had to be leased from the parent or another common carrier on a tariffed basis (Bethesda 1985, II-37).

The other common carriers (GTE, Western Union, and the other independent telephone companies) would be subject to a lesser degree of separation. They were simply required to keep separate books of accounts for their unregulated services. These accounting controls were designed to make it impossible to cross-subsidize competitive activities from revenues produced in the regulated sector where the carrier was guaranteed a favorable rate-of-return. The subsidiaries could only receive basic transmission service from their parent companies on a tariffed basis. In other words, a common carrier would be required to "sell" basic service to its enhanced services subsidiary at the same price, and under the same terms, that it sold it to other enhanced providers. There could be no preferential treatment of the enhanced services subsidiary.

6.5 *COMPUTER III* AND THE MFJ

The ink was hardly dry on the second computer decision when AT&T requested a waiver from the FCC that would have allowed it to provide "custom calling" services without recourse to a separate subsidiary (Kellog, Thorne, and Huber 1992, 550). In regulatory parlance this was referred to as a request for a waiver from "a line of business restriction," that is, a restriction imposed by regulation that prevented a carrier from operating a particular "line" or type of business. One of the services AT&T sought a waiver for was call answering. Call answering is one of the call management services described previously. It is akin to leasing an answering machine from the telephone company. Unlike a home answering machine, however, messages are recorded in a computer operated by the telephone company. The service is more powerful and more flexible than a home answering machine because it allows a subscriber to receive messages while speaking on the phone. The FCC challenged AT&T to demonstrate that there would be unreasonable costs to customers if the service were provided through a separate subsidiary (551). Because AT&T was unable to demonstrate that it was technically impossible to provide the service through a separate subsidiary, the FCC denied the waiver request.

The FCC was faced with numerous other requests for waivers. Although the structural separation requirements were designed to prevent unfair competition, this was not their ultimate purpose. In this case, competition was not an end in itself. Rather it was a means to promote innovation and the introduction of new services. Although a decision not to grant a waiver to AT&T may have been consistent with the *Computer II* rules, this was not a sufficient basis for decision making. The ultimate test of *Computer II* would be whether new services became available to the public. If the FCC denied AT&T's application, and no other firm stepped in to offer the service in its place, then the *Computer II* rules were not meeting their objective.

It is at this point, with the FCC contemplating the efficacy of its *Computer II* rules, that the AT&T divestiture occurred. This had the effect of shifting attention from AT&T to the divested Bell Operating Companies (BOCs), because the BOCs were now the providers of local access and local exchange (switching) service. The reader will recall from chapter five that the agreement that was initially struck between the Department of Justice (DOJ) and AT&T was modified by Judge Harold Greene of the Federal District Court for Washington, D.C. Moreover, Greene determined that he would continue to oversee the implementation of the Modification of Final Judgement (MFJ) in his court. Among other things the MFJ stipulated that the BOCs could not provide "information services." Information services were defined as "the offering of capability for generating, acquiring, storing, transforming, processing, retrieving, utilizing, or making available information which may be conveyed via telecommunications" (*Modification of Final Judgement. United States v. AT&T*, 522 F. (D.D.C. 1982), IV. J). This was a very broad definition. Michael K. Kellog, John Thorne, and Peter W. Huber note that the definition came to encompass such activities as "data processing, electronic publishing, voice answering services, electronic mail, videotex, electronic versions of Yellow Pages directories, E911 emergency service . . ." (1992, 315–316).

The divestiture agreement orchestrated by Attorney General William Baxter was predicated on the principle of total separation. However, this was not the separation of functions or services, it was the separation of *entire companies*. Companies on one side of the divide would operate in deregulated markets (AT&T and Western Electric); companies on the other side would continue to be regulated (the BOCs). Not only did Judge Greene embrace the concept of total separation, it became the philosophical basis for his approach to implementing the final judgement. Judge Greene became a vigilant defender of the boundary instituted by the MFJ—a boundary that relegated the BOCs to the business of basic common carriage at the local level. Following the divestiture, requests by the BOCs for rulings on enhanced services, including waivers from the separate subsidiary requirement (*Computer II*) would be filed with the FCC, and requests for waivers from the prohibition on information services (MFJ) would be filed with Judge Greene. Moreover, the DOJ would continue to play a role in setting policy through its participation in the triennial review of the MFJ.

The two separation frameworks, administered by two different authorities, did little to resolve confusion concerning the role the BOCs should play in the provision of new services. It warrants mention that the restrictions on the provision of enhanced services and information services were not interpreted to mean the same thing. However, over and above the problem of how to interpret the definitions was the problem of whose jurisdiction was preeminent, the court's or the FCC's. Gerald W. Brock has described the conflict of jurisdictions and the confusion it propagated in the following way:

> The FCC's definition of enhanced service and the MFJ's definition of information service were similar but not identical. It was consequently unclear whether the divested BOCs could provide any enhanced services without violating the MFJ prohibition on the provision of information services, but if they could, they were required to do so under an unregulated, fully separate subsidiary. The close relationship between restrictions on the BOC provision of information services, interpreted by Judge Greene, and restrictions on the BOC provision of enhanced services, administered by the FCC, created many opportunities for conflict between the district court and the FCC. Both claimed primacy and suggested the other should yield to avoid conflict, but neither had clear power to overrule the other. (1994, 220)

The conflict between Judge Greene and the FCC was heightened by a shift in policy at the FCC. The slow growth of new services, combined with the difficulties involved in interpreting and enforcing the boundary between basic and enhanced services, convinced the commission that the structural separation requirements were not an effective basis for policy. More specifically, they appeared to be an obstacle to innovation and new services. In 1986 the commission stated:

> We conclude that the record strongly supports a finding that the inefficiencies and other costs to the public associated with structural separation significantly outweigh the corresponding benefits. . . . The relative costs and benefits of the structural separation requirements now imposed on the enhanced services operations of AT&T and the BOC's, compared with the costs and benefits of non-structural safeguards designed to serve the same regulatory goals, lead us to conclude that the structural separation requirements should be eliminated. (FCC 1986 quoted in Brotman 1987, 322–323)

The solution would be a trade-off, a quid pro quo. The BOCs would be allowed to integrate enhanced services with their other operations on the condition that they allow outside providers of enhanced services equal access to their local networks. This would be accomplished in the long-term through a program known as Open Network Architecture (ONA); in the short-term it would be implemented through a framework known as Comparably Efficient Interconnection (CEI).

The new approach was outlined in a number of orders, referred to as the *Computer III* decisions (FCC 1985). The *Computer III* decisions prescribed CEI as a temporary measure, pending approval by the FCC of comprehensive ONA plans that the BOCs were required to file with the commission. The CEI rules stipulated the conditions under which a BOC could provide a *specific* enhanced service on an *unseparated basis*. A specific enhanced service could only be offered by a BOC, as part of an integrated service, if the BOC took measures to ensure that competitive providers of the same service had "equivalent access" to the BOC's network capabilities: The BOC enhanced service and the competitive enhanced service would have to have access to the BOC's local network that was technically and economically equivalent.

The key to long-term policy was the concept of ONA. The term "Open Network Architecture" was coined by the FCC to designate a new way of organizing the relationship, at a technical and organizational level, between local exchange carriers and providers of enhanced services. The key to the new relationship was the "unbundling" of the basic elements that went into the provision of local exchange service. In order to do this the BOCs would have to "fully reveal the technical details of their local exchange networks, including the software in the switches" (Noam 1989, 44). Examples of the elements to be unbundled were to include: dial tone, switching, ring and busy tones, answer supervision, billing, and network management information (Huber 1987, 1.5; Kellog, Thorne, and Huber 1992, 558). Under the initial framework proposed by the FCC, these network functions would comprise Basic Service Elements (BSEs). Once unbundled, these BSEs would be offered to all enhanced providers on a tariffed basis. The tariffs would have to be approved by the FCC. Enhanced service providers could then purchase the unbundled BSEs from the local exchange companies and combine them with additional components as part of a new enhanced service. ONA was conceived by the FCC to be an evolutionary process, not a static set of network functions. It was meant to change and evolve in conjunction with changes in technology, specifically switching technology.

In compliance with FCC requirements, the seven Regional Bell Operating Companies (RBOCs)[2] filed their preliminary ONA plans on February 1, 1988. In November 1988 the FCC approved the plans, but ordered revisions that had to be filed by May 1989 (Bolter, McConnaughey, and Kelsey 1990, 372). On December 22, 1988, the FCC approved the RBOCs' common ONA model. Even so, it ordered more changes. The common model put forward by the RBOCs introduced a new level of service, which they referred to as Basic Serving Arrangements (BSAs). An enhanced service provider had to subscribe to a BSA before it could request an unbundled BSE. For example, Bell Atlantic proposed the following types of network access as part of its BSA:

Voice Grade—Line—Circuit Switched
Voice Grade—Trunk—Circuit Switched
Digital Grade—Circuit Switched
Packet Switched
Dedicated—Private Line[3] (391)

The following are some examples of BSE's proposed by Bell Atlantic:

Calling Number Identification
Call Block
Automatic Call Back
Call Forwarding on Busy Line
Automatic Call Distribution
Closed User Groups (391)

Each RBOC filed a separate ONA plan. The plans differed with respect to the services included as part of the BSAs and the BSEs. For example, Bell Atlantic identified five BSAs, Ameritech four, and NYNEX fourteen (394). The FCC continued to request modifications and amendments to the plans into late 1991. It should be recalled that the ONA approach to enhanced services was part of a regulatory quid pro quo. Once the ONA plans were approved and implemented, the BOCs could begin to offer enhanced services without recourse to a separate subsidiary.

BOC implementation of ONA was subject to considerable controversy. Eli M. Noam has noted that the BOCs' initial enthusiasm for ONA had diminished by the time they filed their first tariffs in 1988 (1989, 80). According to Noam, the BOCs then adopted a containment strategy designed to avoid renting pure switching functions. Whereas the FCC wanted the BOCs to unbundle exchange services into BSEs, the BOCs' plan was based on BSAs. The BSA/BSE model enabled the BOCs to "unbundle the bells and whistles but not the meat and potatoes" (82).

The National Association of Regulatory Commissioners (NARUC)[4] was a harsh critic of the initial BOC ONA plans. In testimony before the U.S. House of Representatives, their representative stated: "As filed, however, the ONA plans are nothing more than warmed-over state tariff filings—they propose little in the way of technological change, and rely instead on repackaging and renaming existing services. It is a regulatory quick fix. This reshuffling of existing tariffs cannot be the technical upheaval required by the FCC's *Computer III* orders to cure the evils of a century of vertical-integration in the telephone industry" (Worthy 1988, 9).

Subsequent filings met with similar dissatisfaction. In February 1991 several industry groups challenged ONA access tariffs filed by Ameritech. Enhanced service providers protested that the tariffs would give Ameritech an unfair advantage (*Intelligent Network News* 1991). The principal complaint concerned the costing model employed by Ameritech to develop its tariff schedule. Long distance carrier MCI challenged the Ameritech tariffs, stating before the FCC:

> With its apparent disregard for commission rules, and the ongoing proceedings attempting to provide an industry consensus to the appropriate access costing modi-

fications necessary for ONA services, Ameritech appears adamant in unilaterally determining what ONA will be in terms of the services offered, and what the access costing and rate making rules will be under its proposal. . . . Ameritech's inability to categorically illustrate its costing methodology and results would open the door for potential strategic pricing of the unbundled access rate elements. (1991)

Another long distance carrier, Allnet, accused Ameritech of " 'unbundling . . . non-cost-based rate elements' and introducing 'new, non-cost-based rate elements' in order to produce 'windfall profits' " (*Intelligent* 1991).

The most significant setback for FCC ONA policy was delivered by a California court on June 6, 1990 (*California v. FCC*, 905 F.2d 1217 (9th Cir. 1990)). The appeals court struck down the FCC's non-structural safeguards and remanded the case back to the FCC for further consideration. The court found that "the record yields no support for the commission's position that market and technological changes since *Computer II* and the BOC Separation Order have reduced the danger of cross-subsidization by the BOCs" (*FCC Week* 1990). The court summarized its position in the following way: "In sum, the commission has failed to explain satisfactorily how changed circumstances justify its substitution of non-structural for structural safeguards to protect telephone ratepayers and enhanced services competitors from cross-subsidization" (1990). In other words, the court recognized the FCC's power to reverse policy and abolish the structural separation safeguards, but such a radical shift in policy had to be explained and justified. Essentially, the FCC had failed to do so. On December 20, 1991, the FCC reinstated the *Computer III* framework including the removal of the requirement for structural separation (FCC 1991). In response to the criticism of the court, the commission strengthened the accounting safeguards designed to prevent cross-subsidies (Kellog, Thorne, and Huber 1992, 581).

While the FCC was implementing *Computer III* based on a quid pro quo that linked the removal of structural separation requirements to the implementation of ONA, the DOJ and the BOCs were moving to challenge the line of business restrictions that had been introduced as part of the MFJ. Recall that the MFJ prohibited the BOCs from engaging in the following businesses: long distance service, the manufacture and sale of equipment, information services, and any other non-telecommunications business subject to regulatory approval. In February 1987 the DOJ issued its first triennial report on the MFJ (U.S. Department of Justice 1987). In this report, issued only five years after the AT&T divestiture, the DOJ reversed its position on the line-of-business restrictions when it recommended that the BOCs be allowed to participate in the following businesses:

1. long distance telephony (outside their monopoly territory)
2. equipment manufacture
3. information services
4. non-telecommunications businesses.

Moreover, the DOJ endorsed the FCC's *Computer III* rules as the most appropriate response to the problem of cross-subsidies in information services. This meant that the BOCs, the DOJ, and the FCC were now aligned together in opposition to the line-of-business restrictions. Opposing them were Judge Greene, AT&T, and other competitors that opposed BOC entry into their markets.

In September 1987 Judge Greene issued a lengthy order that was extremely critical of both the DOJ's new position and the FCC's *Computer III* framework (Order, U.S. District Court for the District of Columbia, Civil Action No. 82-0192. 10 September 1987, "Opinion"). Greene made it clear that his court would continue to enforce the line-of-business restrictions. However, he indicated that he would begin to loosen the restrictions that prevented the BOCs from offering a type of information service known as a gateway.

A gateway is a service that provides users with a single source for locating and gaining access to a wide variety of information services. For example, a gateway typically displays a directory of services that are available to users. It may also act as a billing agent for the various services that are accessed through the gateway. In France, the Direction Générale des Télécommunications, the national telephone company, had created a successful videotex gateway known as Télétel. But in the United States the restriction on information services prevented the BOCs from offering a similar service. On March 7, 1988, Greene amended section VII of the MFJ so that the BOCs could "engage in the transmission of information as part of a gateway to an information service, but not in the generation or manipulation of the content" (Order, U.S. District Court for the District of Columbia, Civil Action No. 82-0192, March 7, 1988, 65). Essentially, the BOCs were allowed to provide an "interface" to the other services, but not the actual services. They would be allowed to organize and display "intermediate" content (information that facilitated access to other services), but not final content.

The DOJ and the BOCs continued their appeal of the general restriction on information services in the courts. In the spring of 1990 the DOJ and the BOCs won their case before the appeals court (*United States v. Western Elec. Co., Inc.*, 900 F.2d 283, 300-305, 309-310 (D.C. Cir. 1990)). The court of appeals determined that Judge Greene had erred when he substituted his own judgement for that of the DOJ and the BOCs. Recall that the DOJ and the BOCs were initially adversaries in the antitrust case, and ultimately, the principal signatories to the MFJ settlement. If the parties, which had been adversaries, came to a new understanding, Greene was obliged to give it due consideration. The appeals court ruled that he had not done so. Also, the court determined that Greene had not used the proper legal standard to interpret the restriction on information services.

Greene had defended the prohibition on information services on the grounds that it was in the "public interest." But the appeals court ruled that the public interest test "directs the district court [that is, Greene's court] to approve an uncontested modification so long as the resulting array of rights and obligations is within the zone of settlements consonant with the public interest today " (*United States v. Western Elec. Co., Inc.*, 900 F.2d 283, 300-305, 309-310 [D.C. Cir. 1990]). Greene had determined that the DOJ/BOC/FCC agreement on information services was not in the best interest of the public. In other words, it was not optimal. But the law did not require that the district court rule on whether an uncontested modification was *the best* possible arrangement; rather it prescribed that it fall within the range of possible solutions compatible with the public interest. On July 21, 1991, Judge Greene issued an opinion and order lifting the information services restriction on the RBOCs. He did so reluctantly, and continued to maintain that the BOC stranglehold over local exchange service would preclude fair competition. Thereafter, the FCC, armed with its *Computer III* framework, would be the arbiter of whether the BOCs would be allowed to provide information services as an integrated part of their network operations.

6.6 THE TELECOMMUNICATIONS ACT OF 1996: LINKING ENHANCED SERVICES TO LOCAL NETWORK COMPETITION

On February 8, 1996, President Bill Clinton signed into law the Telecommunications Act of 1996. The act was the first comprehensive revision of the Communications Act of 1934. The legislative history of the act reveals intense lobbying on the part of the BOCs, long distance carriers, broadcasters, cable companies, and the computing industry. Although less successful, non-profit organizations representing consumers, rural groups, educational interests, libraries, and the health sector also managed to be heard (Aufderheide 1999). The overhaul of communications policy was truly extensive, affecting the telephone, broadcasting, and cable industries. Provisions dealing with obscenity and indecency would have implications for the emerging Internet industry.

The major networks in the broadcasting industry (ABC, NBC, CBS, and Fox) were big winners. Incumbent broadcasters were able to maintain their control of the frequency spectrum for analog television. Moreover, they lobbied hard for, and won, the right to the spectrum that will be used to broadcast the next generation of digital, high-definition TV. The act dramatically loosened controls on concentration of ownership in the broadcasting industry (radio and TV). It is now possible for a company to own radio licenses that cover the entire United States. Moreover, depending on the size of the local market, a company may own up to eight radio stations in a single market. The terrestrial market for broadcast TV will also become more concentrated. For the first time a single company may control licenses providing service to 35 percent of the national market. The ban on cross-ownership of broadcasters and cable operators in the same market has also been eliminated (Huber, Kellog, and Thorne 1996).

The cable industry was another beneficiary of the Telecommunications Act. Price deregulation of everything except the basic tier of service is permitted. Moreover, deregulation of the basic tier will be allowed once a cable operator is faced with competition from an alternative provider of video service in its market. The cable industry is free to enter the market for local phone service, as are the telephone companies to enter the business of video programming.

The act effectively dismantled the 1982 AT&T consent decree, including the prohibition on BOC participation in long distance and the manufacture of telecommunications equipment. For example, the act opened the door for the BOCs to immediately enter the market for long distance in areas *outside* their territory. However, before a BOC (identified in the act as an incumbent local exchange carrier (ILEC)) can enter the market for long distance *within its own territory*, two conditions would have to be met: (1) the FCC would have to determine that the ILEC had complied with a checklist designed to open local networks to competition; and (2) the ILEC would have to secure a binding contract with a competitor for an interconnected service, including, if requested, the provision of unbundled local network functions. When these conditions have been met, the ILEC may provide a competitive long distance service through a *structurally separate* affiliate. After three years the company may apply to the FCC for relief from the separate subsidiary requirement.

The requirements that comprise the checklist are found in section 271 (c) (2) (B) of the act. The checklist begins with interconnection requirements already spelled out in section 251: the obligation to allow resale of services based on wholesale rates, number portability, dialing parity (no extra digits when dialing), access to rights-of-way used by the ILEC, reciprocal compensation, and collocation of equipment (where technically practical). Section 271 goes on to include access to unbundled network elements, access to 911, and access to directory assistance.

Entry into equipment manufacture may occur once the ILEC has met the previously stipulated requirements for entry into long distance. After the ILEC has been in the manufacturing market for three years, the separate subsidiary requirement is removed. Section 274 prohibits the ILECs from engaging *directly* in any electronic publishing venture; these activities must be carried out through a structurally separate affiliate.

The act removes most of the legal and regulatory barriers that have segmented the telephone and cable industries into quarantined markets. But it does this at the expense of reintroducing the vertically integrated provision of services by companies that may combine the manufacture of equipment, the provision of long distance service, the provision of enhanced services, and local exchange operations. However, unlike the monopoly period, the new framework is designed to foster competition between numerous vertically integrated telecommunications companies.

Non-profit organizations were somewhat successful in their attempt to enhance the universal service provisions of the act. The act creates a federal universal service fund and mandates the Joint Board (comprised of federal and state regulators) to identify services eligible for support from the fund. Moreover, the list of eligible services is not fixed, but subject to periodic review by the Joint Board and the FCC. The act stipulates a number of principles that should guide the Joint Board and the FCC in their efforts to design a universal service policy. These principles include the following: quality services should be available at just, reasonable, and affordable rates; access to advanced services should be nation-wide; rural and high-cost areas should have access to basic and advanced services that are equivalent to services available in urban areas; all providers of telecommunications services should be required to contribute to the universal services fund; mechanisms designed to support a universal service policy should be specific and predictable (principle of transparency); and special consideration should be given to ensuring that elementary and secondary schools, libraries, and health care providers will have access to advanced telecommunications services (section 254 (b)).

The act reverses a section of the 1984 Cable Act that prohibited telephone companies from providing video programming. However, should a telephone company enter the video programming business it will be subject to the same regulatory obligations that apply to cable operators, for example, the obligation to reserve one third of its channel capacity for the distribution of local broadcast signals and franchise obligations mandated by local authorities.

The most controversial provision of the act, the one that received the most public attention following its passage, concerned the protection of minors from indecency on the Internet. The amendment, known as the Communications Decency Act (1996), was proposed by Senator James Exon of Nebraska. It made it illegal to engage in indecent communications with a minor over the Internet and to provide access to indecent material via the Internet. It warrants mention that the prohibition was extreme since it went beyond the criminalization of "obscene" communications to ban communications that could be deemed "indecent." It would have left it to prosecutors and the courts to determine what constituted indecent communications. In June 1997 the U.S. Supreme Court ruled that the indecency clause was unconstitutional (*see* Aufderheide 1999, 183–186).

6.7 CONCLUSION

The convergence of computers and communications presented a singular challenge to public utility regulation. Initially, the merging of computers and communications was a technological phenomenon. We have seen that communications have been an integral component of computing since the first commercial computing systems were introduced in the 1950s. However, as the technologies used in communications and data processing began to converge, regulators attempted to keep them apart. Ostensibly, the FCC was not trying separate technologies; rather, it was trying to maintain a boundary for regulatory purposes that had separated two industries: computing and telecommunications. However, it became impossible to maintain the separation between industries without making fine-grained distinctions concerning technologies and their applications.

Under the U.S. Communications Act, the FCC had the authority to regulate the telephone industry, but this mandate was not grounded in the regulation of technology. However, in its zeal to keep the industries separate so that it could continue to exercise regulatory control over the telephone industry, it found itself regulating *on the basis of technology*. This approach had implications for both the industry the FCC was mandated to regulate, and the computer industry, which was outside the purview of its authority. Ultimately, this approach was doomed to failure, since at the very moment the commission was attempting to create a regulatory framework based on distinctions related to technology, applications were being developed that combined the technologies, making them more and more interdependent and indistinguishable. Regulation could only be successful at the expense of limiting the ability of both industries to exploit the inherent potential of hybrid applications to deliver new services and to improve the efficiency of their operations.

The problem for regulators was not simply limited to creating and maintaining industry boundaries. Creating boundaries was merely the first step. The next step was to determine the degree of separation that would be imposed on activities that fell on the "competitive" side of the boundary. The FCC experimented with a number of approaches running the gamut from total prohibition, to maximum structural separation, to protections based on accounting rules designed to prevent cross-subsidies. Underscoring the separations policy was the assumption that the monopolist would not compete fairly if allowed to provide an integrated service that combined monopoly and competitive components. But there was a more fundamental supposition underscoring this policy: It assumed that without competition the monopolist would be slow to innovate and unresponsive to the needs of consumers. So the issue was not simply unfair competition, but the impact of unfair competition on the availability of *new* services.

The case of enhanced services is a striking example of regulatory failure that produced a reversal of regulatory policy. In fact, the court found the FCC's reversal on BOC structural separation to be so sudden and so radical that it sent the new rules back to the FCC for an explicit explanation and justification. The new policy based on ONA and local network competition is grounded in a very different set of assumptions than those that governed the old regime. On the one hand, it assumes that the participation of dominant local carriers in markets for enhanced service is desirable, even essential, to new service development. On the other hand, it assumes that a new type of pro-competitive regulation will be successful in both dismantling local monopolies and ensuring that the local exchange carriers will compete fairly in the market for enhanced service. At a more fundamental level, the new policy assumes that competition in local networks will be both efficient and sustainable.

How does one explain the change in policy? Although there were many contributing factors, two stand out: (1) the FCC's frustration with approaches based on definitions and separations that proved to be unworkable (*Computer I* and *Computer II*); and (2) the FCC's frustration with the slow pace of development of new services. This included frustration with the slow pace of network modernization in general, and, in particular, with disappointing levels of investment in broadband technology that could increase the channel capacity of networks. Broadband networks, based on fiber-optic and coaxial cable, are a prerequisite for many new services that involve high-speed data transmission (computer graphics and video transmission).

Essentially, the FCC and Congress (with the passage of the Telecommunications Act of 1996) have embarked on an ambitious adventure: They have gambled that by allowing BOCs to diversify into previously forbidden services, the BOCs will find incentives to invest in the broadband networks necessary to carry the new services. In effect, what policy makers appear to have done is solve two problems with one policy—the regulatory equivalent of killing two birds with one stone. They have solved the problem of how to promote the development of enhanced services by linking it to the problem of how to open local markets to the forces of competition. In doing so they have cast aside, or deemed irrelevant, over a century of conventional wisdom on the natural monopoly status of local networks. However, it is one thing to change the rules of the game, it is quite another to ensure that the new rules will produce the desired result. Ultimately, the success of pro-competitive regulation will be measured by the availability of enhanced services, the extent of investment in broadband facilities, and the degree of competition that emerges in local markets. If the new rules do not produce the intended result, there will, no doubt, be ample room for questioning whether the rules are to blame—whether policy was deficient in some respect. One wonders, however, whether it will still be possible to question the economic assumptions that underscore the policy, and whether it will still be possible to ponder the question of natural monopoly at the level of local network facilities. Probably not, as this may have to await the next long cycle of regulatory reform.

NOTES

1. A caller provides signalling information to the network when he or she "dials" a telephone number; the switch then processes the digits (signalling information).

2. The seven regional holding companies created by the AT&T divestiture that grouped together the BOCs were: Ameritech, Bell Atlantic, NYNEX, Pacific Telesis, Southwestern Bell, U.S. West, and Bell South.

3. A circuit switched connection is typical of voice telephony where a two-way communications link is maintained between two points (two users). A line connection connects a subscriber to a local exchange (switch). A trunk connection connects two exchanges. Packet switching is used for data communication. Unlike basic voice communication, it does not require a circuit connection, that is, a bi-directional connection for the duration of the communication. Information is codified digitally and converted into series of packets. The packets may follow different paths through a network, but they are reassembled in sequential order at the receiving end.

4. NARUC is an association of state-level regulatory boards throughout the United States.

7

The Canadian Approach to Deregulation

7.1 INTRODUCTION

Pierre Trudeau once remarked that for Canada living next to the United States was comparable to a mouse sleeping alongside an elephant: The mouse had to be very attentive to the elephant's movements lest it turn over in the night. Canada's geographic proximity to the United States, coupled with the fact that the United States is Canada's principal trading partner, meant that Canada was the first country to feel the effects of U.S. deregulation. Limitations of space will prevent us from cataloging all the implications of U.S. deregulation for Canada. Instead we will briefly consider the most important and direct impacts of U.S. deregulation as viewed by Canadian policy makers during the 1970s and 1980s.

The introduction of competition to private-line and public voice services, coupled with the rate rebalancing that accompanied the divestiture of AT&T, had the effect of lowering the price of long distance communications in the United States. This had at least two implications for Canada: higher prices for telecommunications services in Canada would result in higher costs of doing business for Canadian companies when compared to their U.S. counterparts, and the threat of bypass of Canadian telecommunications facilities. The problem was concisely summarized in a report prepared by D. A. Ford Associates for the federal and provincial governments of Canada in 1986. The report noted the following:

> Recent reductions in the rates for U.S. long distance services have increased these differences in rates [i.e., between Canadian and the U.S. carriers]. This situation has caused pressure on Canadian telecommunications regulators and policy-makers in two ways. First, Canadian businesses are aware of the differences in rates, and there is concern for the impact on the competitiveness of Canadian industry. Second, the rate differentials themselves have created perceived economic incentives to avoid Canadian telecommunications carriers in favor of U.S.–based alternatives.

In other words, the difference in long distance rates increases the cost of doing business in Canada compared to the United States, and the difference in rates also causes firms to look for cheaper ways to meet their communications needs, sometimes by "bypassing" Canadian carriers through the use of lower cost U.S.–based service providers. (D. A. Ford and Associates 1986, 1)

Moreover, if Canada were to respond to the differential between Canadian and U.S. long distance rates by lowering the price of long distance in Canada, this would increase the risk of bypass of local networks. This problem was analyzed in a study commissioned by the federal Department of Communications (DOC) in 1984 entitled *The Impact of Bypass on the Future Developments of Local Telecommunications Networks*. Again, the U.S. experience with competition was the impetus for Canadian reflection on the possible negative consequences of competition:

> This study was necessitated as a result of two major happenings. The first is the review of toll competition and rate rebalancing that is currently before the [C]ommission. The second is that the US, as a result of their policy allowing toll competition, have major concerns regarding the impact of local bypass as a direct result of toll competition. The US for some time have been conducting their own studies on the impact of local bypass. In Canada, the potential advent of toll competition and the possibility that toll competition might be a driving force behind local bypass, have resulted in the commissioning of this study. (ii)

Should long distance competition result in reduced traffic on the public, long distance network, this would reduce the contribution (subsidy) that long distance made to the operation of local networks. Ultimately, this would produce rate increases for local service and create incentives for business to bypass the local exchange altogether, which could be done by using private lines to gain access to long distance facilities or to gain direct access to other private networks. As more and more businesses sought alternatives to the local exchange, revenues would decline further, creating a snowball effect. If left unchecked, this would have an effect on common carrier revenues and, ultimately, universality.

Policy makers were also concerned that competition would accelerate the introduction of technological innovations and new services south of the border. Canada would lose its technological edge as a producer of telecommunications equipment and services. Canadian business would find it hard to compete with U.S. corporations who had access to the new services and technology.

To summarize, it was felt that the introduction of competition in the United States threatened the competitiveness of Canadian business in general, and the competitiveness of Canada's telecommunications carriers and equipment manufacturers in particular. If Canadian carriers were unable to match price reductions in the United States, they would probably suffer revenue losses as a result of bypass. The reader will recall from the second chapter how the first transcanadian telegraph services were provided in the nineteenth century using U.S.

facilities—an early case of bypass. If Canada's long distance carriers were to respond to U.S. competition by lowering prices for long distance services in
Canada, this would help Canadian telephone companies to maintain revenues.
But it would threaten the subsidy that funded local service, and by extension,
the principle of universality.

Although the consensus was not universal, it was generally agreed that it
would be beneficial to both consumers and producers if the Canadian response to U.S. deregulation was national in scope (Stanbury 1986). Assuming that some changes to regulation were in the offing, it was clearly preferential from the producers' point of view to deal with a single, uniform,
national approach to competition, rather than multiple regimes specific to
the different provincial regulatory jurisdictions. Compliance to multiple
regimes would add to the costs of providing service and undermine any market economies for services that might exist at a national level. From the consumer's point of view it would be unacceptable for services to be offered in
some parts of the country, as a result of competition, but not in others. This
is not meant to suggest that a consensus existed during this period on the extent to which the Canadian market should be liberalized. This was a separate
issue. However, even from the perspective of companies such as Bell Canada
that opposed competition, it was preferable that if liberalization were to
occur, the same rules would apply throughout the country. Although the problem was how to achieve a coordinated, harmonized set of rules for regulating
the industry, the origin of the problem was U.S. deregulation. Events in the
United States raised the profile of telecommunications as a public issue and
gave urgency to the search for a solution to the problem of national policy.

The solution to this problem did not come as the result of negotiations involving the provinces and Ottawa. It resulted from a Supreme Court decision issued in 1989 known as *Alberta Government Telephones [AGT] and Canadian
Radio-television and Telecommunications Commission (CRTC) and CNCP
Telecommunications (CNCP)* (Supreme Court of Canada, August 14, 1989).
While economists were promoting deregulation to U.S. policy makers, Canadian
policy makers were grappling with a more fundamental problem: how to make
policy at the national level. In the next section we will consider the problem of
national policy, how it dominated the policy agenda of the 1970s and 1980s, and
how the problem was ultimately resolved.

Although the Supreme Court decision dealt with jurisdictional issues, the decision was also an important step on the road to telecommunications liberalization—the principal subject of this chapter. The liberalization of telecommunication markets in Canada can be divided into two periods: pre–Supreme Court
on *AGT* and post–Supreme Court on *AGT* (or pre-decision and post-decision).
In the pre-decision period the CRTC proceeded gradually towards a limited liberalization of markets. In the post-decision period the CRTC accelerated the
pace of liberalization and embraced a policy of open markets.

In subsequent sections we will analyse the actual liberalization and deregulation of specific sectors of the Canadian telecommunications industry: markets for equipment, for long distance, and satellite communications. This sectorial analysis will be followed by a review of three events that had a profound impact on the organization of the Canadian telecommunications industry: the Bell Canada reorganization of 1987, the privatization of Teleglobe in 1987, and the Telecommunications Act of 1993. Although the *AGT* decision was a watershed event, our analysis of Canadian liberalization has not been organized using the pre-decision/post-decision dichotomy. Instead we will focus on specific markets following the model used in the previous chapter on U.S. deregulation. The reader should, nevertheless, be aware that *AGT* had a significant effect on the liberalization of the Canadian telecommunications market. This was particularly true of the process that led to the liberalization of long distance telephony. As we will see in section five, the CRTC rejected an application by Canadian National and Canadian Pacific (CNCP) Telecommunications to enter the market for public, long distance service in the pre-decision period; the commission subsequently reversed itself under different conditions when its jurisdiction had been expanded as a result of *AGT*.

7.2 THE PROBLEM OF NATIONAL POLICY: A UNIQUELY CANADIAN PROBLEM

One of the distinctive features of the Canadian industry structure following the Second World War was the mode of regulation. In Europe and throughout most of the world, a single, national carrier that was owned by the state and operated as a department of government traditionally provided telephone and telegraph services. This department, referred to as a Post, Telegraph, and Telephone (PTT) administration, was responsible for the provision of postal, telegraph, and telephone service throughout the country. In effect, it operated as a monopoly. In the United States there was never public ownership of the telephone industry (table 7.1). But the FCC was able to set national policy through its authority to regulate interstate and international communications.

In Canada, the CRTC was the most important regulator and policy maker for the telecommunications sector. This was the case because Bell Canada and British Columbia Telephone Company (B.C.Tel) Company, who together accounted for just over 70 percent of telephone subscribers in Canada, were regulated by the CRTC (Canadian Radio-television and Telecommunications Commission 1991, 70). Notwithstanding this fact, approximately 30 percent of the Canadian market stood outside the CRTC's jurisdiction. More importantly, in three provinces the dominant telephone carrier was regulated at the provincial level as a result of provincial ownership (Alberta, Saskatchewan,

Table 7.1 Telecommunications Industry/Regulatory Structures: United States, PTT (Europe), and Canada

	Ownership		Regulatory type		National market	Regulatory jurisdiction	
	Private	Public	Independent agency	Department of government	Dominant national carrier	2 tier (split jurisdiction—state and federal)	1 tier (territorially based—provincial or federal)
United States	X		X		X	X	N/A
PTT (Europe)		X		X	X	N/A	N/A
Canada	X	X	X	X*	X	N/A	X

*Where the telephone company was an agency of the provincial government such as Manitoba, Saskatchewan, and Alberta.

and Manitoba). In Nova Scotia, New Brunswick, Prince Edward Island, and Newfoundland, the companies were regulated by independent, provincial regulatory bodies. In Quebec the CRTC regulated Bell Canada, but the province of Quebec, through its Régie des Service Publics, regulated Québec Téléphone and Télébec, the province's second and third largest telephone companies (table 7.2).

After the Second World War, the federal government began to assert its authority more vigorously in a number of jurisdictions. But it was not until the early 1970s that Ottawa began to turn its attention to the telecommunications industry. When government did so in a 1973 Green Paper on telecommunications, it

Table 7.2   Canadian Structure of Divided Regulatory Jurisdiction (1989)

Company	Affiliation	Ownership	Type of Corporation	Regulation
AGT Ltd	Telecom Canada	Public	Crown corporation	Provincial (Alberta)
Bell Canada	Telecom Canada	Private	Investor-owned BCE	Federal
B.C.Tel	Telecom Canada	Private	Investor-owned GTE	Federal
Edmonton Tel.	Telecom Canada	Public	Municipality	Municipal
Island Tel.	Telecom Canada	Private	Investor-owned	Provincial
Manitoba Tel. System	Telecom Canada	Public	Crown corporation	Provincial
Maritime Telegraph & Telephone	Telecom Canada	Private	Investor-owned	Provincial
New Brunswick Telephone	Telecom Canada	Private	Investor-owned Buncor Inc.	Provincial
Newfoundland Telephone	Telecom Canada	Private	Investor-owned	Provincial
NorthwestTel		Private	Investor-owned BCE	Federal
Northern Telephone		Private	Investor-owned	Provincial (Ontario)
Québéc-Téléphone		Private	Investor-owned GTE	Provincial
Sasktel	Telecom Canada	Public	Crown corporation	Provincial
Telebec		Private	Investor-owned	Provincial (Quebec)
Teleglobe		Private	Memotec Data Inc.	Federal
Telesat	Telecom Canada	Private/ Public	Investor+public ownership	Federal
CNCP Telecommunications		Private	Canadian Pacific/ Rogers	Federal

lamented the fact that member companies of the Trans Canada Telephone System (TCTS) were setting policy for the national network with little opportunity for the expression of the public interest on the part of government and other groups (Department of Communications 1973, 7). Moreover, at that time the Canadian Transport Commission (CTC) had no statement of national objectives to guide its decision making. (Responsibility for telecommunications would eventually be transferred to the CRTC in 1976). The closest thing to a policy directive was the statement in the Railway Act of 1908 that tolls had to be fair and reasonable. In the Green Paper the federal government committed itself to enacting new telecommunications legislation that would include a statement of telecommunications policy objectives (13).[1] But the Green Paper did not elaborate what these objectives would be. Instead, it referred to the need for a broad set of objectives for Canadian communications policy, including the broadcasting sector. Canada's communication system (encompassing broadcasting and telecommunications) would be called upon to

1. safeguard, enrich, and strengthen the cultural, political, social, and economic fabric of Canada;
2. contribute to the flow and exchange of regional and cultural information;
3. reflect Canadian identity and the diversity of Canadian cultural and social values;
4. contribute to the development of national unity; and
5. facilitate the orderly development of telecommunications in Canada, and the provision of efficient and economical systems and services at just and reasonable rates. (3)

The Green Paper reaffirmed the principle of Canadian ownership of common carriers and broadcasting undertakings (10), acknowledged that efforts would have to be made to ensure consultation and collaboration between the federal government and the provinces (29), and raised the issue of political (i.e., federal cabinet) control of the CRTC (25). Finally, in the area of common carrier regulation, the Green Paper identified a number of issues that would have to be addressed in future years:

1. How to balance the interests of consumers and producers in the setting of just and reasonable rates
2. The need to create incentives for new services
3. The problem of cross-subsidies between different classes of telecommunications users
4. The need for a policy on the interconnection of telecommunications systems
5. How to deal with the entry of new firms into the common carrier business. (14–17)

In its discussion of common carrier regulation, the federal government hinted for the first time that it was moderately in favor of more competition. In retrospect, we see that the Green Paper was setting the policy agenda for the next two decades; however, change would only be introduced slowly, in measured steps. It would take over two decades for the CRTC to deal with most of these issues. A number of them, such as how to create incentives for investment in new services, still preoccupy policy makers.

The provinces responded to the federal Green Paper by demanding: (1) jurisdiction over all aspects of cable distribution systems (except services engaged in federally regulated broadcasting); (2) the transfer of jurisdiction to regulate all common carriers (except CNCP, Telesat Canada, and Teleglobe); (3) the right to be consulted on the development plans of federally regulated carriers in the areas of standards, frequency management, and satellite competition; and (4) the right to regulate all broadcasting undertakings that were not national in scope (Schultz 1982, 59–60). These hefty demands suggested just how far apart the provinces and the federal government were in their negotiations for a new sharing of powers.

Throughout the 1970s and 1980s attempts were made to negotiate a sharing of powers that would be acceptable to the provinces and Ottawa. Things did not bode well for the discussions when in 1973 the first conference of communication ministers broke up within a few hours (Schultz 1982, 59). Two subsequent conferences held in 1975 were equally unsuccessful. In 1978 a working group on competition and industry structure was created. Although the group was able to reach a consensus on a set of policy objectives and principles, these were not subsequently endorsed by the respective governments of the time; they were merely tabled for further discussion. It was particularly difficult for the governments to reach a consensus on the contentious issue of competition vs. monopoly in the telephone industry (64). In 1987 it appeared that the two levels of government had reached a working consensus on a division of regulatory powers. The *Federal-Provincial-Territorial Memorandum of Understanding*, signed on April 3, 1987, reflected a federal/provincial consensus in two key areas: the need for a consistent national policy on the interconnection of networks, and the need to share and harmonize regulatory responsibility between both levels of government (Federal-Provincial Conference of Communications Ministers 1987). As a result of the memorandum, which became known as the Edmonton Accord, responsibility for regulating the intraprovincial activities of the telephone companies was to be transferred to the provinces. The CRTC would have retained its powers to regulate national carriers such as Teleglobe, Telesat, and CNCP, as well as to regulate the interprovincial activities of the telephone companies (Quebec, ministère des communications 1990, 13). However, the Edmonton Accord was soon overshadowed by the *AGT* decision.

At one level, the jurisdictional conflict in telecommunications must be understood as part of a historical struggle between the provinces and the federal government concerning the division of powers under the Canadian constitution. But there were also a number of substantive disputes specific to the telecommunications sector. These concerned the role of competition and monopoly in the sector, the impact of competition on universality, and the impact of rate rebalancing[2] on the price of basic telephone service. At the centre of these disputes was the CRTC.

By the late 1970s the CRTC had become a cautious advocate of market liberalization. In many respects this was to be expected since the agency could hardly ignore developments south of the border. How the CRTC introduced competition in a series of prudent, incremental steps will be analysed in detail in the following sections. It is sufficient to note that the actions of the CRTC were at the very heart of provincial discontent with the distribution of regulatory powers, adding fuel to the fire of provincial/federal conflict over communications policy. We have chosen two cases of CRTC policy making during this period to illuminate this conflict: the CRTC review of TCTS rate practices and the CRTC policy towards ownership of cable facilities.

One of the most troubling initiatives for the provinces was the CRTC decision in 1978 to launch an extensive inquiry into the rates and practices of the TCTS (later known as Telecom Canada). In its public notice, the CRTC announced that it would look into the fairness of the settlement procedures employed by the TCTS members. The settlement procedures determined how fees for long distance were divided among the members of the TCTS consortium. That is to say, how much of the charge for a long distance telephone call each company would receive when the call was initiated in, terminated in, or transported from its territory. The commission also proposed to examine: (1) whether the rates charged on a cross-Canada basis were just and reasonable; (2) whether the TCTS construction program was reasonable; and (3) whether the TCTS was sufficiently responsive to the demand for the transmission of programming and other information services at a reasonable cost (Schultz 1982, 63).

From the point of view of the Atlantic Provinces, where the telephone companies were largely provincially regulated, and the Prairie Provinces, where the companies were provincially owned, any CRTC interference in rate setting for long distance would have onerous consequences. In the United States competition had led to reductions in rates for long distance service, but rates for local service had increased. Recall that long distance rates have traditionally been set higher than costs in order to subsidize local service. This subsidy was justified as part of a policy to promote universal access to telephone service. Since business is the most important user of long distance, the subsidy constituted an economic transfer (redistribution) from one class of users to another: from business users to residential consumers. Any attempt on the part of the CRTC to follow the U.S. lead in lowering long distance rates through competition would leave the provinces in a difficult situation. This would be the case since long distance competition would create pressure to reduce the long distance subsidy and the burden would fall largely on the provinces to rebalance, that is, increase, local rates in order to compensate for the lost subsidy.

The provinces, and in particular the Prairie Provinces,[3] were cool to the idea of competition and so-called rate rebalancing. They were particularly frustrated by the fact that the CRTC had no mandate, either in legislation or from the federal cabinet, to introduce competition, and there were no formal channels available to the provinces to challenge CRTC policy. Although the federal government may have looked favorably on a more competitive industry structure, there was no official statement of policy at the federal level mandating the CRTC to pursue such a course. From the provincial point of view the CRTC was out of control.

Canada's cable television industry was also caught in the regulatory cross fire. In 1978 the Supreme Court of Canada ruled in favor of federal jurisdiction to regulate cable broadcasting (*La Régie des Service Publics* v. *Dionne* [1978] 2 S.C.R.191; 83 D.L.R.(3d) 181). However, the decision left unresolved the issue of jurisdiction to regulate so-called non-programming services distributed through cable: videotex, security services, metering, and so on. The provinces, and in particular those with telephone companies that were crown corporations, feared that the cable industry would compete with the telephone companies to offer enhanced services. A source of particular irritation for the Prairie Provinces was the CRTC requirement that cable companies own some of the hardware used to provide their service. CRTC regulations required that the cable operator own the receiving antenna, the head-end amplifiers, and the drop to the subscribers (Bruce, Cunard, and Director 1986, 313). In the provinces of Manitoba, Saskatchewan, and Alberta, the hardware (including cables) was owned by the telephone company and leased to the cable operators. The cable companies were regulated by the CRTC; the telephone companies were controlled by the provinces. From the provincial point of view, the CRTC hardware ownership policy was clearly at odds with the type of industry structure they wished to promote. However, since the cable industry was under federal jurisdiction the issue was beyond their purview. Provincial policy was clearly at odds with federal policy.

In 1976 the federal government reached an agreement with the Manitoba government. The agreement stipulated that the "'regulation or supervision of telecommunication services, other than programming services' would be the responsibility of the Province" (Bruce, Cunard, and Director 1986, 313). The agreement appeared to resolve the question of hardware ownership: It stipulated that facilities could be owned by the telephone companies and leased to the cable operators provided they were employed to offer programming services.[4] But in the aftermath of the agreement, the CRTC stated that it did not consider itself bound by the accord and began to issue decisions that did not entirely comply with it (Bruce, Cunard, and Director 1986, 314; Schultz 1982, 62). From the perspective of the provinces, the CRTC was clearly operating outside any effective political control.

Richard Schultz has argued that the record of federal involvement with the telecommunications industry during the 1970s and early 1980s reveals a preoccupation with policies for the *use of telecommunications to further other national objectives*, rather than a willingness to deal with problems of industry structure and other issues fundamental to the development of a strong telecommunications sector (1982, 46–47). The federal government's approach to the cable industry substantiates this argument. Although cable networks have the potential to deliver more than broadcast signals, and, therefore, should be considered in any comprehensive approach to telecommunications development, the federal government's policy towards cable was dominated by *cultural and national unity* concerns associated with broadcasting. Preserving Canadian cultural sovereignty against the incursions of American programming, and preserving the Canadian union against internal division predominated the federal policy agenda. The federal government was singularly reluctant to deal with the issue of the cable industry as a potential distribution network for future telecommunications services.

During the 1970s a number of attempts were made to reform telecommunications legislation. It warrants mention that at least one of these failed attempts, Bill C-16 (1978) included an objective relating to the need for federal/provincial consultation on telecommunications. The bill proposed the following objective: "for the purpose of promoting the orderly development of telecommunications in Canada, there should be consultation between the Minister and the governments of the provinces" (3 (q)). During the 1980s a number of options were discussed to improve provincial input into policy making at the national level. These included regional representation on the CRTC, and more formal mechanisms for consultation and cooperation at the ministerial level (Québec, ministère des communications 1990, 12). As we will see in the next section, the problem of national policy was ultimately resolved in 1989 as the result of a challenge to an application filed before the CRTC in 1982.

7.3 THE SUPREME COURT RULES IN FAVOR OF FEDERAL JURISDICTION: THE STAGE IS SET FOR A NATIONAL APPROACH TO COMPETITION IN TELECOMMUNICATIONS

On September 17, 1982, CNCP Telecommunications filed an application with the CRTC to connect its network with that of Alberta Government Telephones (AGT). If successful, the application would have resulted in a CRTC order obliging AGT to allow CNCP interconnection with AGT's provincial telephone network. Recall that AGT was a publicly owned, provincially controlled telephone company operating in Alberta. It was the largest telephone company operating in the province of Alberta and a member of TCTS.[5]

AGT opposed the application and appealed the case on jurisdictional grounds to the federal court. Essentially, AGT challenged CRTC authority to rule on the interconnection application. In making its case AGT acknowledged that it was a federal undertaking under the meaning of section 92(10(a)) of the British North America (BNA) Act (1867);[6] however, it asserted that it was entitled to immunity from federal regulation because it had been incorporated as a *provincial* crown corporation. Because it was incorporated provincially, AGT argued, it should be considered as an agent of the provincial (not federal) crown. Although AGT initially won its case in federal court, the Federal Court of Appeal reversed the decision. The Court of Appeal determined that AGT had exceeded its statutory mandate when it participated in the interprovincial activities of TCTS (by then known as Telecom Canada). By engaging in interprovincial activities it nullified its right to assert crown immunity from the provisions of the federal Railway Act. (Recall that the Railway Act empowered the CRTC to determine when interconnection is in the public interest and to set the conditions for interconnection.)

In its judgement, delivered on August 14, 1989, the Canadian Supreme Court reversed the appeal court's decision. It ruled that AGT had not voided its crown immunity through its participation in Telecom Canada, thus upholding the initial ruling of Madam Justice Reed of the first level federal court who had recognized AGT's immunity from federal regulation. However, the Supreme Court opened the door to federal regulation of AGT should Parliament enact legislation changing the law (*Alberta Government Telephones [AGT] and Canadian Radio-television and Telecommunications Commission (CRTC) and CNCP Telecommunications (CNCP)* [1989] 98 N.R. 161).

To summarize the ruling, the Supreme Court: (1) accepted the argument that AGT was a federal undertaking and, therefore, subject to Parliament—thus acknowledging CRTC authority to regulate AGT; (2) but it also ruled that AGT was entitled to provincial crown immunity—thereby recognizing that AGT was not subject to the CRTC; and finally, (3) the Supreme Court opened the door to expanded CRTC jurisdiction should changes be introduced to the law.

A key to deciding the first issue was the question of AGT participation in Telecom Canada. The court ruled that the physical location of AGT's facilities was not the determining factor in establishing whether an enterprise was a federal undertaking under section 92(10(a)) of the BNA Act. The crucial issue was the nature of the enterprise itself, and whether there was sufficient "organization interconnection" between AGT and the interprovincial activities of Telecom Canada to warrant a determination that AGT was a *federal* undertaking. When the courts ruled that AGT was a federal undertaking, because of the nature of its activities within Telecom Canada, this implied that the same would hold for the other privately owned members of the Telecom Canada consortium, that is, the Maritime telephone companies. But in AGT's case, the issue was moot because of its legitimate claim to immunity as an agent of the provincial crown.

Where the Supreme Court differed from the Federal Court of Appeal was on the issue of crown immunity. According to the Supreme Court, the interprovincial nature of some of AGT's operations did not effect its provincial crown immunity from federal regulation. However, and this was the decisive component of the decision, the court ruled that the protection afforded by provincial crown status was not immutable. It could be reversed by an act of Parliament should Parliament amend the Railway Act to include a mention of the provincial crown. In other words, if the Parliament of Canada were to amend the Railway Act, extending the jurisdiction of the act to include provincial crown corporations, this would be constitutionally binding. Clearly, if Parliament were to proceed to amend the Railway Act in this way, AGT, Saskatchewan Telecommunications (SaskTel), and the Manitoba Telephone System (all provincial crown corporations) would then come under the regulatory authority of the federal government and its regulatory tribunal, the CRTC. The first part of the ruling suggested that the privately owned carriers of the Atlantic Provinces were already subject to federal authority.

Immediately following the ruling, the privately owned, provincially regulated telephone companies operating in the Atlantic Provinces submitted tariffs to the CRTC for approval. This was an explicit recognition of the new regulatory environment and the CRTC's expanded authority within the new regime. The following companies acknowledged CRTC jurisdiction in this manner: Island Telephone, New Brunswick Telephone, Maritime Telegraph and Telephone, and Newfoundland Telephone. Their decision to submit to CRTC authority was based on the first part of the Supreme Court decision that recognized AGT as a federal undertaking under the meaning of the BNA Act. The companies believed that they would also be considered federal undertakings should the new jurisprudence underscoring the Supreme Court decision be applied to them. Moreover, the companies were privately owned and had never made a claim to crown immunity. This meant that an amendment to the Railway Act would not be required before these companies could be subject to CRTC jurisdiction.

On October 19, 1989, the government introduced an amendment to the Railway Act that would have had the effect of placing the provincially owned and regulated telephone companies of Alberta, Manitoba, and Saskatchewan under federal jurisdiction. However, the amendment died with the adjournment of Parliament.

In 1990 the government of Alberta moved to privatize AGT. A holding company, TELUS Corporation, was created with AGT as its principal subsidiary. The government retained 43 percent of shares in TELUS; the remaining shares were offered to the public on the condition that no individual could hold more than 5 percent (Communications Canada 1992, 18). When AGT was privatized it lost its crown immunity and fell under the jurisdiction of the CRTC. In the wake of the *AGT* decision, SaskTel and the Manitoba Telephone System continued to be provincially regulated, pending the result of negotiations with the federal government.

The Telecommunications Act of 1993 consolidated legislation pertaining to telecommunications in a single statute. It also settled the issue of federal authority to regulate SaskTel and the Manitoba Telephone System. Section 3 of the act states, "This [A]ct is binding on Her Majesty in right of Canada *or a province*" (emphasis added). This statement actualized the parts of the Supreme Court ruling that stated that federal telecommunications law could be made to apply to a company with provincial crown immunity through an explicit assertion of federal authority. As a result of negotiations between Ottawa and the provincial governments of Manitoba and Saskatchewan, it was decided that the act would not apply to the Manitoba Telephone System and SaskTel immediately upon enactment. In the case of Manitoba, the delay in application was a matter of months. Assented to on June 23, 1993, the act came into force on October 25 of the same year. Section 132 of the Telecommunications Act stipulated that the new law would not apply to the Manitoba Telephone System before December 31, 1993. Section 133 provided for a much longer delay in the application of the law in the case of SaskTel: five years after the act came into force, that is, 1998.

The Telecommunications Act was the final act in the constitutional struggle to determine which level of government (provincial or federal) would have the power to regulate Canada's principal common carriers, and by extension, set policy at a national level for the telecommunications sector. In light of the demands made by both federal and provincial actors during the course of two decades of negotiations, how should one interpret the new order that resulted from the Supreme Court decision and the Telecommunications Act? On the one hand, federal authority was confirmed and extended to include responsibility for regulating all of Canada's dominant provincial carriers: the members of the Stentor Canadian Network Management consortium (before 1992 known as Telecom Canada and TCTS). On the other hand, the new regime fell far short of meeting all the demands of the provinces. From the provincial point of view, two types of powers were in dispute: the power to regulate the activities of the provincially based telephone companies, and the power to influence the CRTC and the federal cabinet in matters of nation policy on telecommunications. Recall that at one point the provinces coveted authority to regulate the *intra*provincial operations of carriers operating in their territories. This would have resulted in two-tiered regulation found in the United States. The provinces also lobbied for a reformed CRTC with provincial representation in the commission. Provincial hopes of a two-tiered system died with the Supreme Court decision. However, the revised Broadcasting Act of 1991 included a provision for changes to the CRTC Act in order to ensure regional representation in the CRTC (section 7). As a result, regional members of the commission are appointed from the following regions of Canada: Atlantic Provinces, Quebec, Ontario, Manitoba and Saskatchewan, and British Columbia. A minor concession to provincial demands for federal/provincial consultation was contained in section 13 of the Telecommunications Act. It charged the minister responsible for telecommunications to consult with a designated representative from each of the provinces before making a recommendation to the cabinet with respect to policy in the following areas: general policy; the exemption of a particular class of carriers to the act; the variation, rescission, or referral back to the CRTC of decisions by the cabinet; and orders concerning technical matters. There is no provision in the act for consultation to be binding on the minister, the federal cabinet, or the CRTC.

In the wake of both the *AGT* decision and the Telecommunications Act, the federal government, and by extension its regulatory agent the CRTC, was finally in a decisive position to set national telecommunications communications policy. In section five we will see how these new powers were used to open telecommunications markets. In the next section we will review the CRTC's step-by-step approach to the liberalization of equipment markets, which occurred in the pre-decision period.

7.4 THE LIBERALIZATION OF TERMINAL EQUIPMENT

In a pattern reminiscent of the deregulation of the U.S. equipment market, Bell Canada argued that it was essential to maintain control over telephone sets and other equipment connected to its network if it were to maintain the "integrity" of its network. The company mounted stiff opposition to all attempts to enter this market and liberalize the attachment of foreign terminal equipment to its network. The first challenge to the Canadian terminal equipment monopoly was launched in 1967, more than a decade after the courts had reversed the Federal Communications Commission (FCC) in the *Hush-A-Phone* case.

In an appearance before a House of Commons committee on communications, Dr. H. S. Gellman of DCF Systems[7] urged Parliament to revise Bell's charter in order to oblige the company to draw up rules for the attachment of non-Bell equipment to its network. Parliament did so in 1968. But the revised charter left it to Bell to stipulate the requirements for attaching foreign equipment to its network. The changes to section 5 of the Bell Canada Act (1968) were as follows: "For the protection of the subscribers of the Company and of the public, any equipment, apparatus, line, circuit or device not provided by the company shall only be attached to, connected or interconnected with, or used in connection with the facilities of the Company in conformity with such reasonable requirements as may be prescribed by the Company" (subsection 4). Subsection 5 designated the CTC[8] as the arbiter of the fairness of Bell's requirements. Complaints concerning the reasonableness of Bell's attachment requirements could be brought before the CTC (subsection 6).

It was not until 1975, however, that the revised Bell Charter would be put to the test. In that year Morton Shulman, a member of the Ontario provincial legislature, had his telephone service cut off by Bell. Bell terminated Shulman's telephone service because he was using a telephone accessory called Magicall, which was a device capable of storing several hundred telephone numbers for automatic dialing. Since the changes to its charter had been introduced in 1968, Bell and the other members of the TCTS consortium had begun to allow limited attachment of subscriber owned equipment to their networks. These changes were primarily limited to the market for *data* services. For example, Bell Canada provided a coupler device for use with customer-provided, data terminals (Restrictive Trade Practices Commission 1981, 129). In the case of Magicall, a similar device—which was owned, installed, and maintained by Bell— was already available to Bell subscribers on a rental basis. The company had not gone to the trouble of establishing technical requirements for the attachment of such a device in cases where customers owned it; nor did Bell provide a coupler for such interconnection. Shulman argued before the CTC that Bell provided couplers for other devices that were similar to Magicall and that the CTC should order Bell to provide such a device to him for use with his equipment. The CTC's response to Shulman's appeal revealed the weaknesses of the equipment interconnection provisions of the revised Bell Charter. The CTC determined that it did not have the power to order Bell to respond to Shulman's request for a coupling device. Because Bell had not published any requirements for the attachment of Magicall, the CTC could not rule on the "reasonableness" of those requirements. In other words, it was up to the discretion of Bell to decide which devices it would provide requirements and couplers for (130–131). The next step in the terminal attachment saga would see the issue before the courts.

The Harding Corporation was a distributor of Magicall and several other devices designed to complement equipment offered by Bell. Bell's refusal to provide a coupling device for Magicall hurt sales of the device and prompted Harding to appeal to the CTC for relief. The CTC denied Harding's appeal on the same grounds that it had already refused to intervene in the Shulman case. Harding also distributed a device called Divert-a-Call, which provided a function similar to today's call transfer services. Predictably, Bell threatened to disconnect customers who used this device. On June 5, 1975, Harding took its case to the Quebec Superior Court requesting an injunction against Bell that would prohibit the company from interfering with Harding's clients. One of Harding's clients was the Bank of Montreal. The bank had informed Bell that it was prepared to lease coupling devices from Bell so that it could connect Divert-a-Call to Bell's network. Bell responded that it did not permit Divert-a-Call to be connected to its network; however, it would consider acquiring the equipment itself and leasing it to the bank, which accepted Bell's offer.

Bell's defence before the superior court was based on a challenge to the court's jurisdiction to rule in the matter. According to Bell, the CTC had jurisdiction, not the courts. On October 2, 1975, the court ruled in favor of Harding. Justice Vallerand ruled that the Bank of Montreal and Harding had duly requested Bell to provide requirements for interconnection. Because the judge considered Bell's refusal to be arbitrary, he granted the injunction against Bell. The judge's ruling turned on the issue of Bell's refusal to provide interconnection requirements. Had Bell provided these requirements, Harding and the Bank of Montreal would then have had the option of either accepting them, or, in the case that they found them to be unreasonable, of appealing them before the CTC. In other words, the normal course of appeal was before the CTC. However, since Bell had refused to respond to Harding's request, the case could be brought before the courts. Subsequently, the Quebec Court of Appeal confirmed the superior court's decision. Both courts agreed that section 5(4) of the revised Bell Charter *obliged* the company to provide interconnection requirements. According to the appeal court, the Bell Canada Special Act (1978): "affirms the right of subscribers to attach all apparatus they wish to Bell's telephones: Bell's only right is to prescribe reasonable requirements which appear to it to be imperative 'for the protection of the subscribers of the Company and the public', which it has refused to do" (quoted in Restrictive Trade Practices Commission 1981, 133). Bell took the case to Canada's Supreme Court and lost. The Supreme Court recognized the authority of the courts to interpret Bell's obligations under section 5(4) of the Bell Canada Act (*The Bell Telephone Co. of Canada* v. *Harding Communications Ltd. et al.* (1979) 1 S.C.R. at 403). The *Harding* case resulted in contradictory interpretations of Bell's obligations under the Bell Canada Act. The decision pitted the regulatory agency (the CTC) against the courts, with the regulatory agency favoring the incumbent and the courts favoring the rights of its competitors.

By the time the *Harding* case had cleared the Supreme Court in 1979, jurisdiction over telecommunications had been transferred from the CTC to the CRTC. The next major case to come before a regulatory tribunal and, ultimately before the courts, was initiated the same year the CRTC assumed jurisdiction over telecommunications (1976). Challenge Communications Limited of Toronto sold mobile telephone equipment to business customers. This technology was an early version of today's cellular telephone technology. Initially, these systems required the assistance of a Bell operator in order to connect calls to the public switched network. The next generation technology was automatic, eliminating the need for an operator to make a manual connection. In July 1977 the CRTC approved an application by Bell Canada to offer an automated, mobile telephone service. However, Bell's tariff did not include requirements for the automated interconnection of customer owned equipment. This meant that Challenge Communications could not sell an automated version of its mobile telephone equipment to its customers. In effect, it would not be able to compete with Bell in the market for the new mobile telephone technology. Bell justified its approach by arguing that the older manual systems did not entail a *direct electrical connection* that addressed Bell's network. The old technology required a human operator to make connections. The rationale was not new. Bell had historically drawn the line on interconnection at the point where foreign equipment addressed its network, that is, transmitted signals to Bell's switches and other control equipment. Bell contended that this restriction was necessary in order to protect the integrity of its network.

Challenge brought the case before the CRTC in September 1977, charging that Bell's tariff was discriminatory. Challenge did not base its appeal on section 5 of the Bell Canada Special Act. Rather, it accused Bell of violating section 321.(2) of the Railway Act. Section 321.(2) prohibited common carriers from unjustly discriminating against any person or company in respect of its tolls or any services or facilities it provided. In response to Challenge, the CRTC launched a brief public hearing. In its decision issued in December 1977, the commission agreed with the plaintiff that Bell's behavior was discriminatory. Bell was ordered to revise its tariff to include an option for mobile telephone equipment that was owned and maintained by customers (*Challenge Communications Ltd. v. Bell Canada* (TD CRTC 77-16)). Bell was not successful in its attempts to reverse the decision on appeal to the courts. The *Challenge* case broke important regulatory ground: it was the first time that the non-discriminatory provisions of the Railway Act had been used to protect the rights of *suppliers* of telecommunications services as opposed to end-users (customers) of services.

It would be misleading to suggest that the terminal equipment market was liberalized solely as a result of court challenges and appeals brought before the CRTC. Parallel to the litigious, case-by-case approach, the federal government was working with the common carriers, equipment manufacturers and distributors, and users to develop a cooperative approach. Recall that the principal common carriers under federal jurisdiction in the 1970s were: Bell Canada (Ontario and Quebec), B.C.Tel, and CNCP. The DOC, which was created in 1969, began as early as 1971 to examine the terminal attachment problem. In 1973 the department began to develop a *voluntary* Terminal Attachment Program (TAP) (Restrictive Trade Practices Commission 1981, 151). The cooperative approach was developed in response to the CTC's position that it did not have the power to order the attachment of customer owned terminal equipment. It was discovered that the attachment of foreign equipment was particularly problematic in the area of public-switched services; attachment to private-line services was less of a problem. TAP was introduced in April 1976. In May 1977 a Terminal Attachment Program Advisory Committee (TAPAC) was created. The committee's members represented the following groups: manufacturers, suppliers, common carriers, and users (153).

The TAP program certified equipment and developed technical requirements for the attachment of equipment that did *not* address Bell's network. It did not even consider a wide range of equipment capable of addressing Bell's network, which was the program's principal weakness. Examples of the equipment approved under TAP included the following: plugs and jacks, automatic answering and recording devices, tape recorders, and facsimile equipment (Restrictive Trade Practices Commission 1981, 121). In its report, the Restrictive Trade Practices Commission (RTPC) noted that there was considerable dissatisfaction with the program: "Complaints have been voiced to the effect that TAP moved too slowly, was retrogressive in some respects and imposed overly onerous standards" (153). The principal complaint concerned the failure of TAP to include equipment capable of addressing Bell's network. TAPAC members decided in April 1979 that they would begin to examine technical standards for such equipment. However, Bell and the other common carriers were not enthusiastic participants in this endeavor. Because TAP was a voluntary program, the telephone companies could not be obliged to list interconnection requirements for equipment that addressed the network in their tariffs.

As momentum in favor of liberalized terminal attachment grew, Bell became frustrated with the case-by-case approach. The company turned to the CRTC for a comprehensive ruling. In November 1979 Bell asked the CRTC to determine whether a liberalized market for terminal equipment was in the public interest. Bell did not want the commission to consider the issue of terminal attachment in isolation. The issue of network protection (integrity) was important, but Bell wanted the commission to consider terminal attachment in relation to other issues, such as "the concept of basic service, the impact of liberalized attachment on Bell's operations, the implications for the Canadian manufacturing industry and Bell's competitive position" (Restrictive Trade Practices Commission 1981, 160).

Bell also submitted an interim proposal to modify its attachment rules while it was waiting for the CRTC to rule in the general proceeding. However, Bell's approach fell far short of full liberalization. First, Bell proposed to work with the DOC to establish general standards for all equipment. Second, all equipment would have to be certified by the DOC. Finally, Bell would issue tariffs for certified equipment that would then have to be approved by the CRTC. The Bell plan prescribed a two-level approval process for all new equipment. The equipment would have to be certified by the DOC, and incorporated in a tariff process initiated and controlled largely by Bell. Users would have had to enter into a special agreement with Bell. Rather than proposing a comprehensive tariff, or no tariff at all, Bell was proposing a virtual case-by-case approach that would be slow and bureaucratic. In an interim decision, *Bell Canada—Interim Requirements Regarding the Attachment of Subscriber-Provided Equipment* (TD CRTC 80-13), the commission rejected Bell's proposal. While waiting to rule in the general proceeding, the CRTC established its own interim requirements for the attachment of network-addressing terminal equipment. The following types of equipment were approved for interconnection with Bell's network: equipment that met requirements already published by Bell; equipment of the same class and manufacture as already provided by Bell to its subscribers; and equipment of the same class and manufacture as approved for use in the United States by the FCC (Restrictive Trade Practices Commission 1981, 162). Bell argued unsuccessfully before the CRTC that U.S. standards were inadequate and would undermine the work of the TAPAC.

What happened next illustrates the workings of the appeal mechanism as foreseen in the parliamentary system of government. On August 18, 1980, the minister of transportation and communications for Ontario petitioned the Governor in Council (federal cabinet) to vary Telecom Decision CRTC 80-13 in order to protect Canadian manufacturers of terminal equipment from unfair foreign competition. Specifically, he urged the federal government to ensure that the importation of terminal equipment be based on the principle of reciprocity of international trade. If other countries wished to sell in Canada, then they would have to open their markets to Canadian manufacturers. This followed a statement by the Quebec minister of communications on August 12, 1980, indicating Quebec's dissatisfaction with the CRTC's interim decision. Bell launched its own appeal to cabinet on September 26, 1980 (Restrictive Trade Practices Commission 1981, 163). It took cabinet almost a year to respond to the petition. On May 7, 1981, the federal government announced that the CRTC's interim requirements for terminal equipment would be allowed to stand. In a joint statement the minister of communications and the minister of trade, industry, and commerce noted that TAPAC was in the process of establishing criteria for the attachment of single-line telephones, multi-line telephones, and private branch exchanges (PBXs). The ministers saw no reason to interfere in this process. On the issue of trade reciprocity, the government announced that the two departments would begin negotiations with foreign governments (particularly those of Japan and Europe) with the objective of opening markets in these countries to Canadian manufacturers (165). This was a weak response to the problem of reciprocity, but it was consistent with the U.S. approach, which had done little to protect U.S. manufacturers from a flood of Japanese telephone sets.

On November 23, 1982, the CRTC announced its decision in the comprehensive proceeding on terminal attachment, *Attachment of Subscriber-Provided Terminal Equipment* (TD CRTC 82-14). This decision established the rules for a liberalized regime for terminal attachment for the companies under CRTC jurisdiction. The decision was taken against a background of general dissatisfaction with the manner in which Bell was serving the market for terminal equipment. Before reviewing the decision, we will consider some specific examples of complaints that were registered by suppliers and users at an inquiry undertaken by the RTPC in the early 1980s.

Under Canadian law of the period, the Combines Investigation Act[9] (1970) delegated responsibility for the investigation of monopolistic situations or other activities in restraint of trade to the Director of Investigation and Research (hereafter referred to as the director). (The U.S. equivalent of this law would be the Sherman Antitrust Act [1890] discussed in the previous chapters.) In the late 1950s and early 1960s the director received numerous complaints concerning Bell and Northern Electric Company[10] (Bell's equipment manufacturing subsidiary). The charges against Bell and Northern, which parallelled the charges leveled against AT&T and Western Electric in the United States, alleged that Bell's preferential purchasing of equipment from Northern was unfair to Northern's competitors and had a stifling effect on competition. The director initiated an investigation, and on November 29, 1966, the Royal Canadian Mounted Police raided Bell's head office in Montreal seizing documents pertinent to the case (Rens 1993b, 303). The investigation was protracted; it took the director ten years to conduct the inquiry and write his report. As prescribed in the Combines Investigation Act, the director's report was then submitted to the Restrictive Trade Practices Commission (RTPC).This occurred on December 20, 1976. The RTPC distributed the director's report and began a public hearing process in March 1977. Over two hundred witnesses were heard in hearings conducted across Canada. The final transcripts numbered thirty-five thousand pages (Restrictive Trade Practices Commission 1981, 1). Three reports were issued between 1981 and 1983. One of them dealt with the interconnection of terminal equipment (July 1981). The second dealt with a proposal to reorganize Bell Canada (July 1982). The third examined the problem of vertical integration and its impact on Canada's equipment manufacturing industry.

One of the results of technological innovation and the opening of markets in the United States was the growth of new equipment options for telephone subscribers. A number of consulting firms emerged in Canada to assist companies and government in their choice of equipment. These firms had considerable experience working with users and were ideally placed to evaluate Bell's performance in the terminal market. Consultec Canada Limited was one of the companies providing consulting services to government and business. In testimony before the RTPC the firm asserted that Bell was not able to provide an adequate variety of equipment. The problem, according to Consultec, "arose out of their [Bell's] perceived need to standardize on one or two items of equipment for a particular function in order to gain the economies involved. Standardization resulted in the telcos' lengthening response time and consequently in business firms' lost time and productivity, since the latter require a broad selection of equipment to permit them to find the right machine and right features for each application" (Restrictive Trade Practices Committee 1981, 184).

Telcost Limited, another consulting firm, testified that many businesses were overequipped relative to their needs, while others were underequipped. To illustrate the point the firm drew attention to the case of multi-line telephones. At the time Bell offered five-line telephones and seventeen-line telephones to its subscribers. Nothing was available in the range between five and seventeen lines. As a consequence, many companies whose needs corresponded to the middle range between five and seventeen lines found themselves paying for capacity they could not use. Telcost was also critical of Bell Canada's product line in the area of multi-path intercom services. Bell only offered two-path intercom equipment; if a customer required more than two paths it was obliged to lease a more expensive switchboard system. Finally, Telcost was critical of the equipment that Bell provided for the computerized analysis of traffic data. These systems enable firms to track calling patterns. This information can then be used to assist companies in matching their telecommunications needs with the most appropriate equipment and services. Telcost noted that Bell only provided the *passive* type of traffic data analyser. For companies that had contracted with more than one long distance carrier, the *active* type of device could be programmed to direct calls to the least expensive provider, based on the time of day and other parameters. The active type of traffic analyser was available in the United States, but it was not provided by Bell (Restrictive Trade Practices Committee 1981, 185).

There was no shortage of negative testimony from *suppliers* of terminal equipment. We will limit our review to the evidence presented by suppliers of equipment to the residential and small business market. At the time, a market was just beginning to emerge for subscriber owned equipment: answering machines, recorders, and other devices that did not address the network. Retail outlets such as Radio Shack, the House of Telephones, and the Telephone Store sold equipment manufactured by Northern Telecom (a Bell subsidiary), GTE (the American telephone company that controlled B.C.Tel), and other U.S. manufacturers. In order to use these devices (which were not telephones but telephone accessories), Bell required subscribers to lease a coupler device at a monthly rate of $2.45. The frustrating experience of one retail chain, the House of Telephones, was summarized in the following manner in the RTPC report:

> According to Mr. Monk, these couplers were unnecessary, for he had installed hundreds of telephone answering devices without couplers, against Bell's wishes, without any problems. Moreover, he claimed, when Bell installed a coupler, the telephone answering device rarely worked, and The Telephone Store's repairman would bypass it; thereafter, there would be no pressure from Bell, which had received a fee for installing the coupler and a continuing monthly payment of $2.45 for the useless product. According to Mr. Monk, Bell's coupler-installation service was slow and, in any event, unnecessary, for the customer could have installed the plug himself. (Restrictive Trade Practices Commission 1981, 197)

The House of Telephones and Radio Shack both testified that they had difficulty getting equipment certified by the DOC under TAP. On many occasions equipment that was already approved for use in the United States, failed to pass the Canadian certification process; it then had to be modified (at considerable cost) before it could be sold in Canada. Radio Shack testified that on numerous occasions it was obliged to remove equipment from its product line because of problems encountered with Bell and the other telephone companies (Restrictive Trade Practices Commission 1981, 197–198).

In addition to this unfavorable testimony, the RTPC had received the director's recommendations, as prescribed in the Combines Investigation Act. The director's report was critical of Bell's special relationship with Northern Electric. He urged that Bell be prohibited from the direct sale and installation of terminal equipment to customers. If Bell were to continue to sell equipment, the director proposed, it should be required to do so through a structurally separate subsidiary with a separate book of accounts. Protections would have to be put in place to prevent Bell's monopoly business from subsidizing its equipment business (Restrictive Trade Practices Commission 1981, 200–202).

It is a testimony to the weakness of the Combines Investigation Act, to the power of Bell Canada, and to the inertia in favor of the regulatory status quo, that the RTPC,[11] after six years of work, was unable to make a strong recommendation on what should be done about Bell's vertically integrated monopoly. No doubt, the RTPC was aware that the CRTC was in the process of liberalizing the rules for terminal attachment. This probably played a role in lessening the RTPC's appetite for a solution requiring the dismantling of Bell Canada. As we shall see later in this chapter, Bell Canada was already one step ahead of the regulators. In the early 1980s the company was already considering a plan to reorganize its corporate restructure. The company hoped that by reorganizing its operations it would be able to keep its diversified operations (public telephone, equipment manufacture, Yellow Pages, private networks, and consulting services), while immunizing itself from charges of unfair competition.

When the CRTC's decision on subscriber-provided terminal equipment (TD CRTC 82-14), was announced on November 23, 1982, the event was largely anticlimactic. There were few surprises as the decision continued the approach outlined in the interim decision (TD CRTC 80-13). Two aspects of the commission's decision warrant examination: how the CRTC justified its decision to liberalize, and the manner in which the commission resolved a number of related issues that were crucial to its successful implementation.

Recall that the entire process began as the result of an application brought before the commission by Bell Canada in November 1979. On November 17, 1981, the CRTC decided to transform the proceeding into a general proceeding applicable to all federally regulated, terrestrial carriers: Bell Canada, B.C.Tel, CNCP Telecommunications, NorthwesTel (TD CRTC 82-14, 8). In the application, Bell asked the commission to rule on changes to its rules for terminal attachment by taking into consideration the broad question of whether competition was in the public interest. In other words, it did not want the commission to rule on the narrow grounds that attachment was technically feasible and would not harm Bell's network. Rather it wanted the commission to consider the impact of competition on Bell's entire business: its revenues, and by extension, its ability to use revenues from profitable serv-

ices (such as terminal equipment) to subsidize affordable local telephone
service and the policy of universality.

A number of groups—such as the Canadian Federation of Communications Workers, the National Anti-Poverty Commission, and the Telecommunication Workers Union—supported Bell's contention that competition
would have an adverse financial impact on residential subscribers. Bell and
B.C.Tel argued that they had already suffered revenue losses as a result of the
CRTC's interim decision on terminal attachment. However, a number of parties—such as the province of British Columbia, the Canadian Business
Equipment Manufacturers Association (CBEMA), the Canadian Industrial
Communications Assembly, the Canadian Manufacturers Association
(CMA), and the Consumers' Association of Canada—questioned the validity
of evidence presented in support of this claim. They urged the CRTC to rule
that there was no clear evidence that competition in terminal equipment
would have a significant impact on carrier revenues. The commission could
find no evidence that the quality of telephone service had diminished as a result of the interim rules on terminal attachment (TD CRTC 82-14, 24). On
the issue of trade reciprocity, the commission heard testimony from CBEMA,
CMA, and the director that the opening of the Canadian terminal market
would not impact negatively on Canadian manufacturers. Bell's case was no
doubt hurt when groups representing Canadian manufacturers declared that
they were prepared to meet the challenge of foreign competition.

After considering the evidence, the commission concluded that the potential
benefits to subscribers of competition outweigh the projected disadvantages.
The commission identified the following benefits: "enhanced consumer choice,
both in equipment availability and in the sources of supply; lower prices, as competition encourages each firm to reduce its costs; and, especially for business
subscribers, increased flexibility and efficiency" (TD CRTC 82-14, 29). In short,
competition was in the public interest because it would benefit subscribers in
general, and business users in particular. There was no evidence that competition would seriously cripple or disadvantage Canada's common carriers or manufacturers. On the contrary, it would force Canadian industry to become more
efficient. All equipment manufactured after September 1, 1983, would have to
be certified under TAP before it could be connected to the public switched network (Communications Canada 1992, 19).

Although the interim decision included terminal equipment that addressed
the telephone network, it was limited to extension telephones; a subscriber's
first telephone still had to be provided by the telephone company. Under the
final decision, subscribers were no longer required to lease or purchase their
first telephone from the telephone company. Under the old end-to-end
regime the wiring inside a subscriber's home or business was the property of
the telephone company, which was responsible for maintaining it. Although
telephone companies argued that ownership of inside wiring should be

mandatory for all subscribers, the CRTC rejected this proposal noting that mandatory ownership would raise costs for some users; permitting optional ownership would raise administrative costs. The commission decided that it would deal with the issue of resale and sharing of terminal equipment in a subsequent proceeding. The interconnection of mobile (cellular) telephones would also be considered in a separate hearing. Finally, and most important, the commission dealt with the conditions under which the telephone companies would be allowed to participate in the market for terminal equipment.

Recall that the director had proposed that Bell be required to sell or lease equipment through a structurally separate subsidiary. The CRTC also had the option of requiring the telephone companies to completely divest themselves of their equipment divisions. The commission opted for a moderate approach. It determined that it would be in the public interest for the common carriers to continue to participate in the terminal equipment market at that time. This would be good for consumers as it would enhance choice and promote competition. However, if competition did not emerge, the commission would reconsider carrier participation in the equipment market. At the time, the commission was conducting the third phase of an inquiry into the manner in which the telephone companies identified the costs of their different operations. This was referred to as the *Inquiry into Telecommunications Carriers' Costing and Accounting Procedures: Phase III—Costing of Existing Services* (TD CRTC 85-10). The commission was confident that the *Phase III* proceeding would lead to a method for costing the different components of telephone service—one that would make it impossible for the companies to subsidize competitive operations through revenues generated by the monopoly operations.

7.5 COMPETITION IN LONG DISTANCE: A POLICY FOR CORPORATE USERS

In chapter two we recounted the origins of Canada's domestic long distance duopoly comprising TCTS and CNCP Telecommunications. In the early years, the two groups were differentiated by the fact that the TCTS group provided telephone service and CNCP group provided telegraph service. As the telecommunications industry matured, public telephone service continued to be TCTS's principal business, while Canadian National (CN) Telegraphs and Canadian Pacific (CP) Telegraphs expanded their services directed at the business market to include: Telex, microwave transmission, high-speed data service, and private-line services.[12] The role that the federal government played in promoting this duopoly is illustrated by the case of microwave transmission technology.

In chapter two, we noted that in 1932 Canadian Radio Broadcasting Commission contracted CN Telegraphs and CP Telegraphs to construct a coast-to-coast network for the transmission of radio programming. With the advent

of television in the early 1950s, the Canadian Broadcasting Corporation solicited bids from CN and CP and from TCTS for the construction of a microwave network. Robert E. Babe (1990) notes that the Department of Transportation (which at the time was responsible for licensing radio frequencies)
encouraged the two consortia to form a partnership. Although CN and CP
were amenable to the partnership idea, TCTS was not (129). Led by Bell,
TCTS argued that microwave technology should be developed and owned
solely by the telephone industry. Ultimately, the federal cabinet intervened,
deciding that there would be no monopoly in microwave transmission. By
1958 TCTS had established a national microwave network, while CN and CP
concentrated their network east of Manitoba.

By the mid-1970s the TCTS group and the CNCP group had well established, national networks operating in Canada. CNCP's share of the telecommunications market was only 5.6 percent of the combined gross revenues for
TCTS and CNCP. But in certain markets its share was significantly higher:
computer communications (11.2 percent), telegraph (89.1 percent), broadcast
(13.8 percent), and private-line voice (7 percent) (TD CRTC 79-11, 18–19).
Many of the services such as private-line voice and high-speed data transmission
required interconnection agreements with the member companies of the TCTS
group, that is, Bell Canada, B.C.Tel, AGT, the Maritime Telegraph and Telephone Company, and so on. For example, under the Bell Loop Agreement of
1967: "CNCP undertook to look to Bell as the prime supplier for all its future
requirements for dedicated local distribution facilities and Bell agreed to provide
such facilities within its territory for any purpose other than public message
voice and subject to other specified conditions. CNCP reserved the right . . . to
construct its own local distribution facilities where Bell's construction charges
or timetables were unacceptable" (21).

It became clear to the CNCP group by the mid-1970s that dedicated loops
to their clients' premises were insufficient. In order to remain competitive, a
more flexible interconnection arrangement would have to be secured. Accordingly, Canadian Pacific filed an application before the CRTC for orders
requiring Bell Canada to grant CNCP access to its facilities for the purpose
of interconnecting the two networks. The application was filed on June 14,
1976. Because of the vagueness of the application, it was not initially clear to
either the CRTC or to Bell Canada what services CNCP intended to offer
through the interconnection arrangements. Eventually, it was determined
that the application was for two types of interconnection: Type 1 interconnection involved the connection of the CNCP network to switching equipment in Bell's central offices; Type 2 interconnection was designed to enable
CNCP to access Bell equipment such as computers, concentrators, and multiplexers that were located on a customer's premises.

In Type 1 service, traffic would be routed through Bell central office switches
to trunk lines that would then terminate on CNCP facilities; in Type 2 inter-

connection, trunks owned by CNCP, or leased from Bell, would run from the customer's premises to the Bell central office and from the central office to the CNCP facilities without passing through Bell's local switch. In both cases, Bell would provide the trunk lines at tariffed rates. Type 1 interconnection would enable CNCP to provide *switched* data network services using Bell's local loops. This is also referred to as *dial-up access* since it would allow any user of the public network to dial into CNCP's data network. Type 2 would facilitate: private-line PBX tie trunks, private-line off-premises PBX, and private-line switched networks. A number of services were proposed that could be offered through either Type 1 or Type 2 interconnection: private-line data circuits, private-line PBX foreign exchange, and private-line off-premises Centrex. The application did not entail the provision of a competitive long distance telephone service (also referred to as Message Toll Service [MTS][13]). One of the advantages of Type 2 interconnection was that it would enable a customer to create a backup network in the event of a system failure on the main network provided by Bell.

In considering the application, the CRTC applied a broad public interest test, thus underscoring the view that more was at stake than the technical feasibility of interconnection. Cognisant of the U.S. experience, where ad hoc decision making in the *Execunet* case had resulted in an unplanned liberalization of long distance markets, and a consequent reshaping of the entire telecommunications landscape, the CRTC was at pains to assert that this would not occur as result of its decision. The CRTC acknowledged the dangers involved in creating de facto policy as a result of individual decisions. Because the government had yet to formulate a general policy on competition, the best the CRTC could offer was an assurance that it had proceeded "cautiously with due regard to the broader issues involved" (TD CRTC 79-11, 239).

The CRTC granted CNCP's application in a decision issued on May 17, 1979.[14] In its general conclusion the CRTC observed that competition between CNCP and the TCTS group was an established and desirable feature of the Canadian telecommunications industry. The CRTC concurred with CNCP that the existing loop agreements for dedicated lines were inadequate, and that the absence of an interconnection agreement would lead to "a continuing and significant decline in CNCP's share of the competitive market" (TD CRTC 79-11, 242). In particular, the CRTC found that dial access from the public switched network "would be increasingly important for CNCP's business" (240).

Bell had argued against interconnection on the grounds that competition would reduce economies of scale, thus compromising the efficiency of its operations. The CRTC found the evidence in support of economies of scale inconclusive. But it did accept the argument that the local exchange facilities of the public network exhibited characteristics of natural monopoly. On the basis of this, it concluded that any duplication of the network at this level would "involve a staggering economic and social cost and would be clearly against the public interest" (TD CRTC 79-11, 240). In general, the commission found that the services proposed by CNCP would provide significant benefits to users, particularly as competition would have the desirable effect of improving the responsiveness of telephone companies to the needs of the market. Type 2 interconnection would give users more options and greater flexibility "to mix and match facilities as they require as well as for back-up purposes" (241).

Most important, the commission rejected Bell's estimates of future revenue losses due to competition. Bell submitted evidence suggesting that it would forego $235.3 million in revenues as a result of competition; the commission was of the opinion that less than $45.7 million per year was at stake. As a matter for speculation, it felt that the shortfall could be made up with minimal increases of $0.47 per month for residential and $2.05 per month for business subscribers. Moreover, the commission suggested that most of the shortfall could be made up among the class of subscribers who would benefit the most, that is, business users.

In granting the application the commission imposed a number of terms and conditions. Limitations were placed on the types of services that could be offered through interconnection: no Wide Area Telephone Service (WATS)[15] or MTS. Measures to protect the technical integrity of Bell's network would have to be implemented. CNCP would be required to compensate Bell Canada in a fair and expedient manner for access to its facilities. Finally, mechanisms would have to be put in place to monitor the effects and availability of the new services.

In spite of CNCP's assurances that it would not exploit interconnection to offer services analogous to MTS or WATS (i.e., a public telephone service), the commission instituted a number of controls designed to reduce the opportunities for abuse. The following limitations were imposed: (1) the service was to be limited to the subscriber's private communications needs; (2) one end of the transmission circuit had to terminate at the subscriber's premises; (3) transmission circuits owned by CNCP, which used connections to Bell's network, had to be dedicated to the subscriber's private use; and (4) in Type 1 service, CNCP was interdicted from introducing its own analogue switch between its interconnection point with Bell and the customer's premises (TD CRTC 79-11, 247).

In rendering its decision, the CRTC drew the line on facilities competition at the boundary between public and private *voice* service. The decision opened the way to competition in public data networks and it expanded opportunities for competition in private-line services. By drawing the line where it did, the CRTC deferred consideration of the contentious issue of rate rebalancing and its attendant debate on pricing principles (cost-based vs. value-of-service/system-wide price averaging). Although CNCP made it clear that it was not seeking to compete in long distance telephony, the CRTC felt it incumbent upon itself not only to protect WATS and MTS from competition, but to justify its policy. It began by asserting that there was no basis in statute for Bell's monopoly in MTS and WATS. Rather it was on the basis of public interest considerations that protection was required. Three justifications were given for constraining competition in public telephony: (1) the potential erosion of MTS–WATS revenues; (2) the significant contribution that MTS made to the costs of Bell's local exchange facilities; and (3) the desirability of maintaining a uniform route-averaging MTS rate structure (TD CRTC 79-11, 245). In justifying its decision in these terms, the commission appeared to be endorsing Bell's contention that MTS subsidized local exchange to a significant extent. But a closer reading reveals that the commission's enthusiasm for the cross-subsidy hypothesis was tempered by an awareness that Bell's other interexchange services, such as private line and data communications, were probably not making an adequate contribution relative to their use of local exchange facilities (223).

Four years later, on October 25, 1983, CNCP dropped the other shoe. The company filed an application with the CRTC to allow interconnection with Bell's network for the purpose of offering MTS and WATS services. Not content with dial-up access to its data network, CNCP was now requesting interconnection to offer a public voice network. The commission followed the same procedure in considering this application as it had done with the earlier application in 1976. It accepted on a prima facie basis CNCP's claim that "interconnection would be useful to its business, that the duplication of the facilities sought would not be desirable in the public interest, and that no unreasonable technical harm would result from the interconnection" (TD CRTC 85-19). Accordingly, the burden of proof was on Bell Canada, the incumbent, to demonstrate that it would not be in the public interest to grant CNCP's application. Again the commission was prepared to consider the effects of interconnection on a broad range of matters that included the following:

1. Universality of service
2. Consumer choice and responsiveness to consumer need
3. Quality of service
4. The justness and reasonableness of subscriber rates
5. The requirement that rates and conditions of service not confer an undue preference or disadvantage
6. Innovation in the telecommunications industry, and the Canadian business generally
7. Efficiency of telecommunications systems
8. Optimal allocation of resources taking account of geographic differences
9. The structure of rates, including route-averaged pricing, rate group structures, and rural service rates
10. Industry structure (11)

Although the CNCP application was rejected in a decision issued August 29, 1985, the commission acknowledged a number of potential benefits of competition. Competition would increase pressure to lower MTS and WATS rates, bringing them closer to costs. Competition would improve productivity, thus lowering costs for all carriers. There would be greater consumer choice and supplier responsiveness to consumers needs. Competition would promote a more rapid introduction of new switching and transmission facilities (TD CRTC 85-19, 44). However, the commission determined that these benefits would not be present to a significant extent throughout the territories served by Bell (44). Most important, the commission was not convinced that CNCP would be able to meet the objectives that it had set out in its business plan. In its ruling the commission stated: "If the [C]ommission granted CNCP's application and required it to make contribution payments equal to those made by Bell and B.C.Tel, CNCP would be financially restricted to offering very limited price discounts and serving a limited number of routes. This would have the effect that a number of the benefits of competition noted earlier would be substantially reduced" (46).

Ultimately, the commission's decision came down to the issue of how much CNCP should contribute to the costs of operating the local exchange service provided by Bell. Because the margin for profit proposed in CNCP's business plan was so small, the commission feared that CNCP would have no alternative but to return at a later date and request a contribution discount. In other words, CNCP would return to the commission and argue that it was unable to contribute to the operation of the local network at the same level as Bell's long distance service. The commission rejected the option of a contribution discount on the grounds that this would constitute an unfair benefit to CNCP subscribers. With foresight the commission did not wish to find its hands tied should CNCP return at a later date, once its business was operating, to renegotiate a lower, more favorable level of contribution.

In conjunction with the hearing, the commission reviewed a key peripheral issue associated with a more competitive industry structure: the problem of rate rebalancing. Rate rebalancing was high on the agenda of Bell Canada and B.C.Tel. In the evidence they presented to the commission, the telephone companies did not attempt to make a strong case for natural monopoly in long distance telephony. Instead they argued that the cost-price differential associated with rate setting in traditional rate-of-return regulation had placed them at an unfair disadvantage; if left uncorrected it would likely foster forms of uneconomic bypass. (Uneconomic bypass occurs when a competitor is able to offer a service at a lower price than an incumbent because regulation has inflated the price of the incumbent's service; the price differential between the incumbent and his or her competitor is not a result of more efficient operation by the competitor, rather it is a function of the obligations imposed on the incumbent by the regulator.) To support their claims both Bell Canada and B.C.Tel presented evidence delineating the cost-revenue relationship for different types of services. The results were presented in a modified version of a model referred to as the Five-Way Split. The model was modified to accommodate an estimate of costs associated with a new accounting category, "access," that had emerged as a result of U.S. deregulation.

Access costs are those associated with the installation and maintenance of the physical plant (facilities) *between a customer's premises and the first telephone switch*. These costs are fixed and do not vary with traffic on the network. The same cost is incurred whether the subscriber uses the telephone once a day or several dozen times a day. This is not true of the switching component of local service where higher volumes of individual use result in an increase in aggregate traffic and may require the installation of more powerful switches. (In practical terms, when a switch is overloaded the subscriber will not get a dial tone.) As long as the network is operated as a monopoly there is little reason to differentiate between local exchange (switching) and access service. They are simply the costs of providing public telephone service at the local level. Moreover, access service is essential to the provision of long distance service since calls can neither be initiated nor completed without it. Bell contended that providing local service to a subscriber in the Ottawa area cost the company $24 per month, for which it only received $9.35 in subscriber revenues (Babe 1990, 142). Using Bell's data from 1983, the CRTC compiled its own table of revenues and costs associated with the provision of public telephone service.

The commission's version of the Five-Way Split demonstrated that both local switched service and long distance (toll) service were highly profitable. Access service was operating at a deficit, but this was really an accounting problem. According to the commission, Bell was able to show a deficit in local exchange service by allocating *all* local access costs to the local service category. But there was no reason to allocate them exclusively to local exchange when long distance also made a claim on access service. Based on the results of the modified Five-Way Split the commission concluded: "neither the competitive network nor the competitive terminal category make any substantial contribution to the recovery of access category costs, that for the local and toll categories, revenues are in excess of costs and that the surplus of toll revenues over costs is the major contributor to the recovery of access costs" (TD CRTC 85-19, 54).

The commission estimated that a fully cost-based approach to rate setting based on Bell's model would entail a more than doubling of local exchange rates, while long distance rates would decrease by 50 percent. While acknowledging that cost-based pricing would reduce incentives for bypass, the commission did not concur with the telephone companies that bypass represented a significant threat to their businesses for the immediate future. The CRTC acknowledged that opportunities for bypass might increase over the long-term, thus exerting pressure for a lowering of MTS–WATS rates, but it did not conclude from this that cost-based pricing would necessarily be the appropriate response. Instead, the commission set in motion a process designed to reduce MTS–WATS prices at a rate equivalent to the rate of inflation. Although such a program would result in some increases in rates for local exchange service, the commission did not feel that these would be of such a magnitude that they would threaten the principle of universality (TD CRTC 85-19, 69).

The CRTC was charting a prudent course on the issue of rate rebalancing for local and long distance service. From the commission's point of view the crucial issue was the competitiveness of Canadian long distance service, not a new formula for allocating costs between the different services. The commission encouraged Bell and B.C.Tel to reduce the price of MTS and WATS. In exchange, the commission was prepared to oversee some increases in rates for local service, and most important, to protect the MTS–WATS market from competition.

CNCP waited five years (May 16, 1990) before it returned to the CRTC with another application to compete with Bell, and the other members of Telecom Canada, in the markets for MTS and WATS. By this time interexchange competition had spread to the United Kingdom and Japan, and the dust had settled from the U.S. experiment with rate rebalancing and the dismantling of the AT&T monopoly. As noted previously, the *AGT* decision expanded the scope of CRTC authority to include the Atlantic Provinces, and potentially all of the provinces. This enabled CNCP to argue in its business plan that it would be able to offer a truly *national service*. This was not the case in its previous application. In the previous application this was an important handicap that figured prominently in the commission's decision to refuse the application. Also CNCP came to the second application in a stronger position as the result of a corporate reorganization. In 1988 Canadian Pacific purchased Canadian National's telecommunications shares. Then in September 1989 Rogers Communications acquired a 40 percent equity interest in Canadian Pacific Telecommunications. The Rogers–Canadian Pacific partnership resulted in the formation of Unitel Communications, which became the new applicant before the CRTC (fig. 7.1). Through its subsidiary, Rogers Cablesystems, Rogers Communications owned Canada's largest cable provider, serving 1.8 million subscribers through fourteen cable companies (Rogers Communications 1991, 1). Finally, and no doubt most significant, Unitel had the backing of some of Canada's largest corporations who viewed competition as a long overdue opportunity to reduce telecommunications costs. Although corporate users were generally supportive of CNCP's first application, their efforts were poorly coordinated and would pale in comparision to the lobbying efforts they mounted in the second long distance campaign.

It was not long after the first long distance application had been rejected that academic economists and lawyers in Canada began preparing for the next round. Two policy institutes, the Institute for Research on Public Policy (IRPP) and the Fraser Institute, financed studies on future telecommunications that were favorable to competition (Stanbury 1986; Globerman 1988). The IRPP studies emphasized developments in the United States suggesting that a combination of rapid technological change, the demonstration effect of U.S. deregulation, and the potential for bypass of Canadian facilities, would

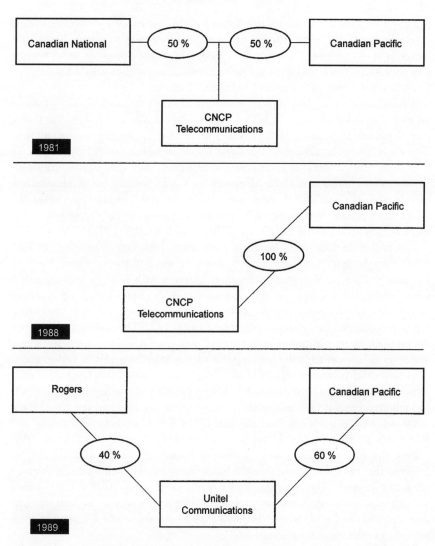

Figure 7.1   From CN and CP to Unitel: The Evolution of Ownership of Canada's Second National Carrier

make competition inevitable. There would be winners and losers: corporations would be the big winners and residential users would be the primary losers (Janisch 1986). However, according to the IRPP studies, the challenge for policy would not be to control the extent of gains to business users, but rather to manage loses to residential subscribers. In these early studies there was no mistaking or masking the fact that competition policy was policy for the corporate sector. If competition resulted in low-income residential users dropping out of the network, the proposed solution was to target direct subsidies at these groups (Globerman 1988). The higher price of local service was simply the price that the average residential subscriber would have to pay to ensure the competitiveness of Canadian industry.

A number of umbrella groups representing business interests were instrumental in the push for change: the Business Council on National Issues, the Canadian Bankers Association (CBA), the Information Technology Association of Canada (ITAC), the Canadian Business Telecommunications Alliance (CBTA), Canadians for Competitive Telecommunications, and the Communications Competition Coalition (CCC). Moreover, as Vanda Rideout has demonstrated, membership in these organizations was interconnected. The membership of the CCC, for example, was composed primarily of multinational corporations that were already active in the CBTA, the ITAC, and the CBA (1997).

The corporate user revolt began in earnest in 1989, four years after the first CNCP application had been turned down. It was led by the Royal Bank of Canada, Bell's most important corporate customer with yearly telecommunications costs exceeding $100 million. Lawrence Surtees notes that in the summer of 1989 the Royal Bank "brought about twenty select corporate users and influential pro-competition warriors together" (1994, 192). The secret meeting produced the Communications Competition Coalition (CCC), a single-issue group whose purpose was to lobby for long distance competition. Allan Taylor, chairman and chief executive officer of the Royal Bank, broke with the tradition of informal, high-level negotiations with Bell, in order to take center stage in efforts to mobilize corporate user support for competition. To this end the Royal Bank commissioned a study by Hudson Janisch of the University of Toronto and Richard Schultz of McGill University. The study entitled, *Exploiting the Information Revolution: Telecommunications Issues and Options for Canada* (1989), became the basis for Taylor's public indictment of the Bell Canada/Telecom Canada monopoly. The corporate users' revolt set the stage for CNCP–Unitel's second application. As Canada entered the 1990s it was clear to both the government and the CRTC where the business lobby stood on the question of competition in telecommunications.

Unitel's long distance application was more sophisticated, more comprehensive, and more theoretical than that of its predecessor, CNCP. The cornerstone of Unitel's application was the claim, detailed in its business plan, that it could offer an average 15 percent discount to consumers, while at the same time making a fair contribution to the operation of the incumbent's local exchange service. The application contained a good measure of speculation on the *ancillary benefits* of competition. Unitel claimed that as a result of competition the market for long distance would grow with respect to both the volume of traffic and the variety of services. In effect, competition would increase the size of the long distance market. As a result of competition, all common carriers (both incumbents and entrants) would attain higher levels of productivity. The overall efficiency of the telecommunications industry would improve. There would be more innovation and a greater responsiveness to the needs of consumers on the part of all carriers. The realization of these ancillary benefits of competition was fundamental to Unitel's business plan. In other words, the application was not just about lower rates (in the order of 15 percent) but rather, it was about the benefits of competition as a total package. In evaluating the application the CRTC was obliged to consider whether Unitel's projections for market growth and greater productivity for the whole industry were reasonable. This made the hearing process somewhat more theoretical and speculative than is generally the case in applications before the commission.

Recall that in the previous CRTC decision (TD CRTC 85-19) the commission was principally concerned with the problem of how to lower rates for MTS and WATS. In that decision the commission acknowledged the textbook benefits of competition, but it did not consider them to be sufficiently important to grant CNCP's application. On June 12, 1992, when it issued its decision in the second long distance application, the commission reversed its priorities: the benefits of competition had taken center stage, and rate reductions were secondary to a new, more competitive industry structure (*Competition in the Provision of Public Long Distance Voice Telephone Services and Related Resale and Sharing Issues*, TD CRTC 92-12). This shift is evident when we consider the commission's treatment of the "Reference Plans" for rate reductions in long distance submitted by Bell Canada and B.C.Tel.

During the period 1986 to 1990, Bell's long distance costs had declined 10.5 percent while B.C.Tel's had been reduced by 8.4 percent. This was in response to the commission's prodding in the previous decision (TD CRTC 85-19, 69). Bell came to the Unitel proceedings with a reference plan that detailed how the company would match U.S. rates for medium- and large-volume users by 1996. Reductions for residential users would come later. By the year 2002 Bell proposed average reductions in toll (MTS–WATS) rates of 55 percent (TD CRTC 92-12, 22). B.C.Tel presented a similar proposal. Moreover, Bell and B.C.Tel did not make these reductions contingent upon rate increases for local service. In 1985 this would have been enough to forestall the threat of competition. In 1992 it was not enough. The rules of the game had changed. The commission explained its decision to favor competition in the following terms: "The [C]ommission considers that more benefits would be achieved if the respondents [i.e., the incumbent telephone companies] were to implement their pricing strategies, but in a more competitive environment. Competition would not only provide a strong incentive for the respondents to continue to minimize costs, but would also offer users a level of choice and supplier responsiveness that cannot be replicated in a facilities-based monopoly environment" (36).

The Unitel application was the occasion for one of the most comprehensive proceedings in the CRTC's history. There were two applicants (Unitel and

British Columbia Rail Telecommunications/Lightel [BCRL][16]), six respondents
(incumbent telephone companies), and over eighty-five official interveners. Approximately two thousand letters and submissions were sent to the commission
by individual citizens, municipalities, and other organizations (TD CRTC 92-12,
5). Public consultation included hearings in each province and territory of
Canada. From the time Unitel submitted its application until the commission
ruled on June 12, 1992, over two years had passed. The printed proceedings,
which could fill a small library, are probably the most thoroughly documented
case on record of a national inquiry into the implications of toll competition for
the telecommunications industry and its subscribers.

Notwithstanding the tremendous scale of the proceedings, the decision to
grant the application came down to a very basic, almost circular, argument: (1)
competition is synonymous with innovation, choice, and responsiveness; (2) regulated monopoly cannot meet these criteria to the same degree as a competitive
industry structure; (3) the time had come to introduce greater innovation,
choice, and responsiveness to the Canadian telephone industry; and (4) the time
had come for competition.

Bud Sherman, the CRTC's vice president for telecommunications, was the
principal architect of the CRTC decision. Four of the five CRTC commissioners who participated in the proceedings shared Sherman's vision; the fifth commissioner, Edward Ross, dissented. In an interview with Surtees, Sherman explained his support for competition in terms of the "power of choice" (1994,
274). Surtees quotes Sherman as saying, "For me, the strong thing was the compelling case that was made that choice would benefit users" (274). It was important that Unitel present a credible business plan; "The math had to be there,
as well" (274). But for Sherman the decision to sanction competition turned on
the question of *consumer choice*. By definition, "choice and innovation" are not
strong selling points for regulated monopoly. Regulated monopoly is valued in
circumstances where a priority is given to efficiency, stability, affordability, fairness, *and the needs of residential subscribers*. It is difficult to imagine how under
these circumstances, and given these changed priorities, Bell and B.C.Tel could
have argued in favor of monopoly. Of course, the emphasis on choice and innovation was largely a rhetorical smoke sceen. What was really at stake was lower
costs to Canada's corporate users. The mantra of choice and innovation for all
users, corporate and residential, had more to do with the CRTC's marketing of
the decision than it did with actual priorities of the commission with regard to
the class of residential users.

It is important to note that Bell had already begun to lose its public relations case with the class of residential users well before Unitel presented its
second long distance application in 1990. In 1986, shortly after CNCP's first
long distance application had been refused, the CRTC ruled that Bell
Canada's profits were excessive. In Telecom Decision CRTC 86-17, *Bell
Canada—Review of Revenue Requirements for the Years 1985, 1986 and 1987*,

Bell was ordered to refund $206 million to its customers. This refund was equivalent to offering each of Bell's more than six million customers a credit for two months of local service. Moreover, the company was ordered to lower long distance rates by an average of 20 percent. The decision was a watershed event. Until that time it was generally believed that rates for long distance could only be reduced in conjunction with either competition or rate rebalancing. When the CRTC ordered Bell to refund its revenue surplus to subscribers and lower long distance rates to prevent further surpluses, it was demonstrating that the price of long distance could be reduced without affecting the price of local service (Surtees 1994, 148).

Bell and the other members of Telecom Canada had also lost ground on another front: the resale of private-line services. Although the commission refused CNCP's application in 1985 to provide competitive MTS and WATS (TD CRTC 85-19), it left the door open in that decision to the eventual *resale* of private lines to third parties for commercial purposes. In a similar vein, the commission announced that it would be open to the *sharing* of private lines. Resale exists when a provider of telecommunications services leases private lines from a facilities-based carrier (such as Bell, B.C.Tel, or CNCP) at a bulk rate, and then makes them available to third parties on a commercial basis. The reseller's service usually entails adding value to the leased private lines. For example, the reseller may enhance the transmission service by providing detailed billing or traffic information that the facilities provider does not offer. Because the reseller leases and does not own the basic transmission facilities, he or she is referred to as a *non-facilities-based provider*. An opportunity for profit exists in resale because of the price differential between public toll service (MTS and WATS), and the discounted rate for private lines. However, under normal conditions the price differential may not be sufficient to draw users to the reseller. When the price differential is combined with extra features designed to meet the specialized needs of a client, the reseller's service may become commercially viable. The market niche for resale services is small- and medium-sized businesses. Leasing private lines directly from facilities providers is not economical for these businesses because the volume of their telecommunications traffic is too low.

In Telecom Decision CRTC 85-19 (which denied CNCP's first application to provide MTS–WATS), the CRTC acknowledged the benefits to consumers of resale and sharing. However, it was not prepared to sanction resale and sharing at that time because it considered the price differential between rates for MTS–WATS and private lines to be too great. This would have given resellers an unfair advantage vis-à-vis the facilities carriers responsible for MTS–WATS. The solution was to wait for MTS–WATS prices to come down, as foreseen in the decision. When this occurred, viable competition would be established on the basis of the resellers' ability to *add value* to basic transmission and not because of an artificial price differential created by regulation.

In Telecom Decision CRTC 87-2, *Tariff Revisions to Resale and Sharing*, the commission opened the market for resale and sharing with some restrictions. The principle limitation was the requirement that there would be no *joint use* of circuits provided by resellers to their customers. Each circuit on a private line that had been leased by a reseller from a facilities provider had to be assigned to a single client. The commission stated:

> With regard to resale, the [C]ommission agrees with Bell's and B.C.Tel's proposed approach. The [C]ommission considers that the provision of services in the nature of private-line voice services entails a relationship in which the user of the service leases a fixed number of circuits for a fixed amount of time. Further, the [C]ommission considers that joint use of reseller-provided circuits by the reseller's customers, as would be allowed under CNCP's proposed restrictions, could result in services in the nature of MTS/WATS that would result in substantial MTS/WATS contribution erosion. (23)

In the case of sharing, where a group of *users* contracted directly with a facilities-based carrier for the joint use of leased-lines, the restriction would not apply. The restriction on joint use was imposed because the commission feared that these services would lead to a substantial erosion of MTS–WATS revenues. This would reduce the contribution that public toll service made to local access service, thus putting at risk the policy goal of universality.

On March 1, 1990, the CRTC removed the restriction on the joint use of circuits provided by resellers (*Resale and Sharing of Private Line Services*, TD CRTC 90-3). The change in policy was made at the urging of the Canadian Federation of Independent Business (CFIB), which argued that the restriction on joint use discriminated against the class of small business users. The CFIB noted that the CRTC had already allowed revenue erosion to occur as a result of the liberalization of the private-line market; large business users were the principal beneficiaries of this policy. It was, therefore, unfair and discriminatory for the commission to deny small businesses access to discounted services on the basis that it might lead to contribution erosion. Furthermore, when the commission looked at estimates of market loss due to a removal of the restriction on joint use, it concluded that Bell and B.C.Tel could make up for the loss with a monthly local rate increase of around fifty cents (II, C). But the commission was not convinced that this would be necessary, since the resale and sharing of private lines could also lead to an expanded market that would increase the overall amount of contribution payments, including payments made by resellers to the telephone companies, as prescribed in the decision. The decision stipulated that a charge of $200 per channel per month had to be paid by resellers to the telephone companies. The commission considered this to be a fair contribution to local access service.

A study of the resale industry by D. A. Ford & Associates released in May 1992, shortly before the decision to liberalize long distance, found that thirty firms had been registered with the CRTC as resale providers. Resellers were concentrated in Ontario, Quebec, and British Columbia, where the CRTC had long-standing jurisdiction. The annual gross revenues of the resale industry were approximately $200 million as of December 1991 (Ford 1992, 5). This represented about 2 percent of the long distance market only two years after Telecom Decision CRTC 90-3 (Surtees 1994, 272). Seventy percent of traffic was within Canada. U.S.–Canadian traffic accounted for 25 percent of the resellers' business. The remaining 5 percent was international (Ford 1992, 6). Small business was the principal market for resellers with 81 percent of customers having monthly bills of less than $1,500 (15). The discounts offered by resellers were in the following range:

- 15 percent to 40 percent discounts on direct dialed long distance
- 40 percent to 60 percent on foreign exchange and off premise extensions
- 20 percent on Dataroute data circuits
- 10 percent on Megaroute data circuits (14)

Bell and B.C.Tel had no doubt anticipated that the CRTC would open the long distance market to competition. The portents of change were already evident: the refund decision (based on excessive revenues), the resale decisions, the commission's generally favorable view of competition in Telecom Decision CRTC 85-19, and the demonstration effect of competition in the United States. Only a month before the CRTC decision, the president of Bell Canada Enterprises (BCE) had announced that Bell would not appeal a negative ruling (Surtees 1994, 270). Nevertheless, the terms of the decision came as shock to Bell Canada.

Surtees has described Bell's astonished reaction in the following terms: "Bell Canada was stunned by the depth of its defeat. 'The worst-case scenario we envisaged happened,' says George Hariton, a member of Bell's team. 'Jean [Monty] and Brian [Hewat] went into major shock.' Although Bell's experts had envisaged open entry, says Hariton, they had not predicted the terms and conditions to be so favorable to competitors" (279).

Bell's frustration stemmed from the financial terms of the decision that it viewed as extremely favorable to Unitel. The decision granted Unitel a discount on the contribution it would pay to local access during the first five years of operation. The initial discount was 25 percent; this would be reduced in subsequent years until it reached zero after the fifth year. Moreover, the commission mandated that the telephone companies would have to bear a significant portion of the costs of adapting their networks for competition: $160 million. Bell was so exasperated by the decision that it reversed itself on its pledge not to appeal. On July 6, 1992, the federal court agreed to hear Bell's charge that the decision constituted an illegal expropriation of the property rights of Bell Canada without compensation (Surtees 1994, 288). In conjunction with the legal appeal, Bell announced plans to further reduce rates for long distance. In August 1992 Bell filed a petition with the federal cabinet, urging the government to reduce regulation of long distance (292). From the CRTC's point of view, this was a blatant attempt by Bell to stifle competition before it was even underway. Surtees argues that Bell's vigorous opposition to the CRTC decision only served to "alienate customers and to aggravate both federal regulators and politicians" (295).

The CRTC decision was clearly a triumph for Unitel. But it was an even greater triumph for Canada's corporate users as the CRTC opened the door to competition wider than anticipated. The decision included approval of an application by a joint venture comprising B.C. Rail Telecommunications and Lightel Incorporated (BCRL). BCRL had applied to construct a fiber-optic network along Canadian National Railway's right-of-way from Toronto to Fort Erie, where it would connect with U.S. carriers in Buffalo (Surtees 1994, 206). Unitel had based its business plan on a scenario involving competition between two facilities-based competitors: Unitel and Stentor (Telecom Canada). (In January 1992 Telecom Canada was reorganized and given a new name: Stentor Canadian Network Management.) Not only did the commission approve BCRL's application to construct transmission facilities, but it announced that it was prepared to consider applications by other facilities providers—on the condition that they were prepared to play by the same set of rules as Unitel and BCRL. Moreover, the decision had the effect of immediately extending the resale market to the Atlantic Provinces. Unitel would face competition from both Stentor and the resellers. Unitel's initial plan called for a roll out of its service to residential users one year after the approval of its application. Potential competition from resellers forced Unitel to reconsider its plans: the company decided to begin residential service in the fall of 1992. On November 4, 1992, Unitel began providing service to eleven thousand customers in Montreal[17] (312).

The final chapter in the early story of long distance competition concerns the search for U.S. partners. Finding a U.S. partner had always been a priority for Unitel. There were obvious financial, marketing, and technical reasons for this. Unitel needed foreign equity to finance its infrastructure investment. The company had limited experience in selling services to the residential market and in the technical platforms needed to provide service to this market. However, it was the Stentor Group, led by Bell Canada, that announced the first partnership agreement with a U.S. long distance carrier.

Although Bell had a long corporate history of close dealings with AT&T, Stentor announced on September 10, 1992, that it had formed a strategic alliance with MCI. The announcement occurred before Unitel had inaugurated its service and had the effect of drawing Unitel and AT&T closer together. Why did Stentor turn to MCI rather than AT&T? It appears that AT&T wanted more control in a partnership arrangement than Bell Canada was prepared to cede. Also, AT&T wanted to employ switching platforms manufactured by its own equipment manufacturing branch, Western Electric. For obvious reasons, Bell Canada favored equipment manufactured by Northern Telecom. Moreover, MCI was already using Northern Telecom switching platforms. A final irritant was Bell Canada's decision in March 1992 to participate in the Financial Network Association (FNA). The association was initiated by MCI as a global alliance of telecommunications carriers to provide services to the financial industry (Surtees 1994, 268–269). FNA was created to compete with AT&T in the financial services market.

On January 7, 1993, Unitel and AT&T announced that they had formed a partnership. The details of the formal agreement were negotiated in the weeks that followed and announced on January 29. AT&T would receive a 20 percent equity stake in Unitel and have two seats on Unitel's nine-member Board of Directors. Unitel would receive two AT&T switching platforms valued at $150 million. To make room for AT&T, Canadian Pacific's share of Unitel equity would be reduced from 60 to 48 percent; Roger's holdings would be reduced from 40 to 32 percent. Canadian restrictions on foreign ownership of telecommunications facilities limited AT&T's holdings to 20 percent of an operating company. Shortly after the agreement was signed AT&T moved to assume management control of Unitel:

> Unitel's employees soon had little doubt who they were working for when AT&T was asked to provide the company with top management expertise. Within several weeks, most of Unitel's top executive positions were taken over by four senior AT&T executives. AT&T officials were quick to point out that their arrival was at the request of shareholders and that their presence was only temporary—for five years. Yet many employees and industry watchers saw it as a corporate coup and, at the very least, a vote of non-confidence in Unitel's top management. (Surtees 1994, 322)

Call-Net Telecommunications, Canada's largest reseller was also obliged to seek a U.S. partner. Call-Net was founded in 1986 to take advantage of the CRTC's new rules on resale. In 1993 Call-Net formed an alliance with Sprint Communications Company of the United States—the third largest U.S. long distance carrier. Under the terms of the agreement, Sprint United States received a 25 percent non-voting equity stake in Call-Net Enterprises. In exchange, Call-Net would be able to use the Sprint name, and have access to Sprint's network, technology, and marketing programs (table 7.3). Sprint Canada, Canada's third largest long distance provider, is a wholly owned subsidiary of Call-Net Enterprises of Canada. Lightel Corporation, one of the partners in the BCRL venture is a unit of Call-Net Enterprises.

In 1997 the total value of the Canadian long distance market was approximately Can $8 billion. Call-Net Enterprises, the reseller whose service is known under the name Sprint Canada, has been the principal beneficiary of competition. Predictably, Bell and the other Stentor members have watched their market share fall below 70 percent. In 1997 Sprint Canada and Unitel (now known as AT&T Canada) each held 8 percent of the market (Riga 1997,

Table 7.3 Strategic Alliances between Canadian and U.S. Long Distance Carriers

	1992	1993	1993
Canada	Bell Canada	Unitel	Call-Net
	↑↓	↑↓	↑↓
United States	MCI	AT&T	Sprint

F-4). The other resellers held the rest of the market. The other principal resellers included Fonorola, ACT Long Distance, and CAM Net.

Before concluding this section on long distance, it is useful to consider how Unitel faired during the early years of the competitive regime it spearheaded. It is relevant to consider Unitel's profitability because it is the key indicator of the sustainability of *facilities-based* competition in the Canadian market. Recall from chapter three that the *un*sustainability of competition is one of the conditions associated with the existence of natural monopoly. In April 1995 (two and half years after it had entered the long distance market) it was reported that Unitel was losing $1 million a day; moreover, a loan in the amount of $659 million that it had secured from Canadian banks was scheduled to mature at the end of the month (Rubin 1995, D-1). The heavy loses prompted Unitel's Canadian partners (Rogers Communications and Canadian Pacific) to seek relief. The first Canadian partner to abandon ship was Canadian Pacific, when it offered to sell its stake to Rogers in the spring of 1995. When Rogers failed to exercise its option, the banks consented to extend Unitel's credit on the condition that a consultant be hired to look into a reorganization of Unitel's ownership. Ultimately, an independent committee was formed that worked through the summer of 1995 on a proposal to restructure Unitel's ownership. On September 26, 1995, a plan was announced allowing Canadian Pacific and Rogers Communication to liquidate their ownership of the money-losing company. The new owners were AT&T and the Canadian banks that held the Unitel loans: Bank of Nova Scotia, Toronto-Dominion Bank, and the Royal Bank of Canada. On January 31, 1996, Unitel's new ownership restructure was announced. On September 9, 1996, it was announced that Unitel Communications Company would change its name to AT&T Canada Long Distance Services Company (AT&T Canada). In what could only appear as a deserving irony, the same banks that spearheaded the drive for competition in telecommunications found themselves in the dubious position of being the principal shareholders of AT&T Canada, Canada's money-losing long distance carrier.

The Telecommunications Act (1993) stipulates that a telecommunications carrier operating in Canada must be Canadian owned and controlled (section 16). The shareholder reorganization announced in January 1996 raised concerns about the degree of management control that AT&T would exercise over the operations of Unitel. Following the shareholder reorganization, Call-Net and Stentor, in separate interventions before the CRTC, requested that

Unitel's ownership be reviewed. Stentor argued that the CRTC should conduct a public proceeding to "consider and disclose all facts and circumstances relevant to 'control in fact' of Unitel, and to adopt policy guidelines regarding control of Canadian carriers" (Darling 1996). The commission rejected the idea of holding a public hearing.

In a letter to Janet Yale, senior vice president of AT&T Canada, Allan J. Darling, secretary general of the CRTC, announced the commission's findings with respect to the question of AT&T's control of Unitel (Darling 1996). The Telecommunications Act limits foreign control of a *holding company* to 33⅓ percent. AT&T Canada (ATC) controls 33 percent of voting shares in UCHI (Unitel Communications Holding Incorporated). ATC is a wholly owned subsidiary of AT&T Corporation of the United States. The remaining voting shares of UCHI are controlled by three banks: Bank of Nova Scotia (27.9 percent), Toronto-Dominion (22.9 percent), and Royal Bank of Canada (16.2 percent). The Board of Directors of UCHI comprises nine members. Under the shareholder agreement approved January 31, 1996, ATC appoints three directors; the banks appoint three members; and three independent members are jointly appointed by the banks and ATC. (This was a departure from the initial agreement whereby Canadian Pacific appointed four directors, Rogers appointed three directors, and ATC appointed two directors.) Essentially, the commission found that the company was not in violation of Canadian law with respect to foreign ownership. AT&T's 33 percent voting interest in UCHI was within the foreign ownership limits established by Canadian law for a holding company.

Although the commission did not find the company in violation of Canadian law with respect to the question of ownership control of Unitel, it warrants mention that the extent of AT&T's actual interest in the Canadian carrier is hardly reflected in the 33 percent measure of voting shares in UCHI. AT&T's actual *economic* interest in Unitel would appear to be closer to 50 percent. What follows illustrates how a complex web of corporate relationships involving holding companies, subsidiaries, voting shares, and non-voting shares can be woven together in order to circumvent regulation and isolate companies from direct scrutiny.

Unitel Communications Company (UCC) is the actual operating company, the telecommunications carrier. (UCC is known publicly as AT&T Canada.) Unitel Communications Holding Incorporated (comprising the banks and ATC) does not exercise direct control over UCC, rather it does so through a wholly owned *subsidiary*, known as UCSP. UCSP holds 50 percent of *non-voting* common shares of UCC, the operating company. The other 50 percent of *non-voting* common shares of UCC is held by AT&T Canada L. D. Holdings (AT&T Holdco). AT&T Holdco is a wholly owned subsidiary of AT&T Corporation of the United States. Therefore, AT&T Corporation (U.S.), through its subsidiary ATC, controls 33 percent of the *voting shares* of the holding company UCHI, and through its other subsidiary, AT&T Holdco, it controls 50 percent of the *non-voting common shares* in UCC, the operating company. Through its voting shares, AT&T participates in the selection of the Board of Directors of UCHI, which in turn is responsible for choosing senior management at UCC and setting corporate strategy. Through its non-voting shares it has a 50 percent economic interest in the company's actual performance.

Five years after the liberalization of MTS–WATS, the Canadian experience with competition could be summed up in the following phrase: competition had been much better to users, and in particular to corporate users, than it had to providers. But the final chapter has yet to be written. Ultimately, the success of the CRTC's experiment with competition will depend on the ability of the long distance companies to turn a profit. Bell and Stentor have lost a significant share of the long distance market; but Bell continues to remain profitable due to its diversified activities. Five years after liberalization the other providers (resellers and AT&T Canada) had yet to turn an annual profit. Most problematic is the status of AT&T Canada, the former Unitel. Because facilities competition is an important part of the competitive equation, AT&T Canada's ability to compete in the market for long distance will be a key to a competitive industry structure. However, if Bell and the other Stentor members dominate the facilities market, regulation, not competition, will set the price of these facilities. The CRTC may once again find itself regulating the market for a key sector of the telecommunications industry.

7.6 CANADA'S DOMESTIC SATELLITE COMMUNICATION INDUSTRY: THE TELESAT DILEMMA

Canada's satellite communication industry comprises two sectors: international and domestic. As was the case in the United States, Canada's participation in international satellite communications preceded its involvement in domestic satellite services. Canada's international, overseas carrier, the Canadian Overseas Telecommunications Corporation (COTC),[18] was Canada's representative in the International Telecommunications Satellite (Intelsat) consortium founded in 1964.[19] In this section we will consider the history of the *domestic* satellite industry in Canada. More specifically, we will consider how the CRTC attempted to introduce competition to the domestic satellite market, and how these efforts were frustrated by both the federal cabinet and the member companies of the Telecom Canada consortium.

Canada's domestic satellite industry was officially born with the passage of the Telesat Canada Act on June 27, 1969. The act created Telesat Canada as a "mixed corporation"[20] with the government holding 49 percent of shares, Bell Canada holding 24 percent of shares, other TCTS Canada members holding 16 percent of shares, other telecommunications carriers with 9 percent of shares, and Telesat employees owning 2 percent of shares (Janisch 1992, 18).

When the act was passed, government ownership of Telesat was viewed as temporary: when Telesat became a viable commercial enterprise, the government planned to sell its shares to the public. Because this did not occur, the government was obliged to retain its shares. Telesat's lack of profitability had a profound impact on Canadian domestic satellite policy during the 1970s and 1980s. The Telesat experience is a relevant and instructive case study in Canadian telecommunications policy because it illustrates the difficulties inherent in policy making for a country such as Canada whose national telecommunications market is small (compared to the United States) and dominated by powerful incumbent carriers.

The reader will recall the U.S. open skies policy that promoted competition by restricting AT&T–COMSAT participation in the domestic satellite market. The FCC was eager to license new carriers; but in order to protect them from potential unfair competition by AT&T and COMSAT, it gave new entrants a three-year head start in the market for competitive services. The reader will also recall that the U.S. domestic satellite system was largely underutilised until a national market for cable TV programming emerged. The Canadian model, embodied in the Telesat Canada Act (1969), rejected intermodal competition. (The term "intermodal competition" refers to competition between incumbent *terrestrial* carriers and *in space* [satellite] carriers.) That is to say, it rejected a fully competitive industry structure that would have seen new satellite carriers compete directly with existing terrestrial carriers. Instead, the Canadian approach sought a *partnership* between the incumbent terrestrial carriers (primarily Bell Canada and the other members of the TCTS consortium [later Telecom Canada]) and the newly created satellite carrier, Telesat Canada. The Canadian compromise created Telesat as a *carrier's carrier*. This meant that Telesat could only sell channel capacity to existing, terrestrial common carriers; it could not market its services directly to large corporations.

Underscoring and justifying the partnership approach was the supposition that Canada's terrestrial common carriers would have to cooperate with Telesat if the satellite carrier were to succeed. Even under a competitive model, which the government rejected, their cooperation would have been required to ensure local connections. When Telesat was created, very little was known about the market for specialized business services via satellite. Given unproven demand for satellite services, the limited size of the Canadian market, and strong TCTS opposition to competition by satellites, the government attempted to garner support for a Canadian domestic satellite project by making the terrestrial common carriers stakeholders in the venture. However, this approach was doomed to failure since it placed the TCTS members in a conflict of interest. As shareholders in Telesat, they had a vested interest in Telesat's success; but this conflicted with their existing and principal interest as operators of terrestrial facilities. There was no incentive for the incumbent carriers to use satellite capacity, as this would draw traffic away from their terrestrial networks where they had already invested and where their revenues (under the rate-of-return formula) were in large part determined by the extent of this investment. The Telesat compromise placed Bell Canada and the other common carriers exactly where they wanted to be: as shareholders in Telesat, they were able to control a potential competitor to their terrestrial networks. Moreover, because they were the only customers authorized to lease Telesat circuits, they were able to determine the level of use of satellite facilities. Because its facilities were underutilised, Telesat was unprofitable. It was not long before Telesat was faced with its first crisis.

In the mid-1970s, not long after the launch of the first *Anik* satellites, Telesat began to plan for an upgrade of its satellites to second generation technology capable of transmitting in the 14/12 GHz band[21] (Bruce, Cunard, and Director 1986, 316). The problem was how to finance the new technology. Because Telesat was not profitable, it would have been futile to attempt to raise capital on the stock market through an offering of shares. Moreover, the government still retained control of the shares that it had initially intended to sell. Telesat could not ask the government for an infusion of capital, because the Telesat Canada Act had stipulated that total government investment in Telesat could not exceed $30 million (Telesat Canada Act 1968-69, c. 51, s.1). The proposed solution was for Telesat to become a member of the TCTS consortium. If this were to occur, Bell and the other TCTS members would then provide additional financing. It was also suggested that the inclusion of Telesat in the TCTS group would facilitate the integration of Telesat's facilities with those of Canada's principal terrestrial carriers. The government approved Telesat joining TCTS on the condition that the CRTC review the new arrangement.

In Telecom Decision CRTC 77-10, *Telesat, Canada Proposed Agreement with Trans-Canada Telephone System,* issued August 24, 1977, the CRTC rejected the agreement that would have made Telesat a member of TCTS. The commission determined that the deal was not in the public interest because the agreement, "would make it difficult to regulate rates, might make it likely that TCTS members would be afforded an undue preference with respect to satellite services and, significantly, that the [A]greement might lessen any nascent competition between carriers" (Bruce, Cunard, and Director 1986, 317). (The reader will recall that CNCP Telecommunications was not a member of TCTS.) The CRTC was clearly hoping that Telesat would eventually blossom into a full-fledged competitor to TCTS in the long distance market. But the government was primarily concerned with the problem of financing Telesat modernization without having to resort to public funds.

On November 3, 1977, the government issued an Order in Council (P.C. 1977-3152) that varied the CRTC decision. Robert R. Bruce, Jeffrey P. Cunard, and Mark D. Director have summarized the government's position on the potential benefits of Telesat joining TCTS in the following manner:

First, significant financial resources would immediately be made available to Telesat. Second, the [A]greement contemplated a system of transfer payments to Telesat in initial years and from Telesat to TCTS in subsequent years, with a guaranteed rate-of-return to Telesat equal to that of B.C.Tel and Bell Canada. Third, the transfer payment system created an incentive for TCTS members to use Telesat's capacity. Fourth, membership in TCTS would give Telesat a more solid financial footing with which it could, with government support, pursue its public policy goals of serving the North and fostering the space industry. (1986, 317)

In 1981 the CRTC again attempted to dismantle the ties that bound Telesat to TCTS. This time it was on the occasion of an application by Bell Canada for increases in the tolls for services it offered in conjunction with TCTS. The application prompted the CRTC to initiate a general hearing on TCTS rates, practices, and procedures, *Bell Canada, British Columbia Telephone Company and Telesat Canada: Increases and Decreases in Rates for Services and Facilities Furnished on a Canada-Wide Basis by Members of the TransCanada Telephone System, and Related Matters* (TD CRTC 81-13, 4). Because Telesat was by this time a member of TCTS, its tariffs were included in the review.

In the course of the proceedings a number of interveners, most notably the National Bank, protested Telesat's exclusive dealings with TCTS. They argued that Telesat's refusal to deal with companies that were not regulated, common carriers violated paragraph 321 of the Railway Act, which stipulated that it was illegal for a carrier to give undue or unreasonable preference to a person or discriminate against any person. The CRTC agreed with the National Bank and ordered Telesat to remove the tariff limitation that restricted end-users to common carriers (TD CRTC 81-13, V.4.a). Also the commission ruled that Telesat's practice of only leasing full channels discriminated against small carriers and ordered Telesat to change its policy (V.4.b). Once again the federal government intervened to vary a CRTC decision concerning Telesat. Through an Order in Council issued December 8, 1981, the federal cabinet decided to maintain Telesat's status as a carrier's carrier. However, an exception was made with respect to dealings with broadcasters such as cable TV networks and the Canadian Broadcasting Corporation (CBC). Telesat would be allowed to enter into contracts directly with Canada's broadcasters, but it could not deal directly with other types of businesses (Order in Council P.C. 1981-3456).

Eventually, the government would open the satellite market. This was done gradually, not by the CRTC, but through the DOC's authority to license the earth stations (antennae) required to transmit and receive satellite signals. Between 1973 and 1979 all earth stations were owned and operated exclusively by Telesat (Bruce, Cunard, and Director 1986, 319). Beginning in 1979 the DOC authorized broadcasters to own "receive-only earth stations" and common carriers to own "transmit stations" for 14/12 GHz services. Then in 1983 the DOC allowed individuals and small businesses to own receiving antennae without a license. Finally, in April 1984 the DOC announced that it would it would license transmitting stations, beginning with non-carrier business users, for use in private-line services. By April 1, 1986, there would be full liberalization of earth stations.

By 1992 a Communications Canada publication noted that Telesat was leasing satellite capacity to "broadcasters, cable television operators, other Telecom Canada members and smaller firms that provide telecommunication services to business" (15). The same publication stated that Telesat was providing services directly to business.

Some examples of Telesat diversification into new markets are the services the company offers through a partnership with Canadian Satellite Communications (Cancom). In 1981 Cancom received its first license from the CRTC to rebroadcast television programming using channels provided by Telesat. The company was formed as a strategic alliance involving four independent broadcasters: BCTV (Vancouver), CITV (Edmonton), CHCH (Hamilton), and CFTM (Montreal). It began by broadcasting the signals of the four partners; in 1997 Cancom broadcast twenty-one channels. These included signals from American broadcasters such as ABC, CBS, NBC, FOX, and PBS. In 1995 the company broadcast programming to 2,600 small cable operators throughout Canada, reaching 3.6 million homes (Canadian Satellite Communications 1995). In the late 1980s Cancom began providing a data transmission service to Canada's trucking industry. The service enables shipping companies to track the movement of trucks across the continent.

In 1992 the Canadian government finally sold its 49 percent share in Telesat for $155 million to Alouette Telecommunications, a strategic alliance of Canadian telephone companies and Spar Aerospace (Lindgren 1997, D3). BCE controls 58.7 percent of Telesat shares through Alouette Telecommunications. However, its voting interest is limited to 26.1 percent (Bell Canada Enterprises 1997). By 1997 Telesat had successfully launched eleven satellites. In 1997 it was Canada's only operator of domestic satellites providing wireless data networks used for "point-of-sale, electronic banking and credit card verification, airline and travel reservations, retail inventory management, video conferencing, distance education and business television" (Telesat Canada 1997). But Telesat's principal business is still the distribution of programming within Canada and across the U.S. border.

Nearly three decades after its creation, Telesat still had a monopoly over satellite communication in Canada. Moreover, the company was more firmly controlled than ever by Canada's domestic telephone companies, principally by BCE and the other members of the Stentor consortium. Under the new ownership arrangement, Canada's traditional terrestrial carriers have retained effective control over Telesat. On February 15, 1997, an agreement on basic telecommunications was concluded at the World Trade Organization. Canada was a signatory to the agreement. Among other things the agreement provided for an end to the Telesat monopoly on fixed satellite services on March 1, 2000 (Industry Canada 1997).

7.7 THE BELL CANADA REORGANIZATION

In section five we saw how AT&T's entry into the Canadian market prompted changes in Unitel's ownership structure. The new structure was designed to better reflect the heightened role that AT&T would play in the financing and management of the company, while respecting the limitations on foreign ownership of Canadian carriers prescribed in section 16 of the Telecommunications Act (*see* the next section). In the case of Unitel, it was the problematic nature of the carrier's foreign ownership that dictated the elaborate structure of holding companies and subsidiaries. Bell Canada has also experimented with corporate forms but to a different purpose. Beginning in the mid-1960s, Bell sought to diversify its operations. But the provisions of its charter, which restricted the company's operations to the telephone business, stood in the way.

Recall that Bell's original charter,[22] voted by Parliament on April 28, 1880, empowered the company to manufacture telephone equipment and provide telephone service to *all* regions of Canada. The company could construct its lines along public roads and all public rights-of-way, subject to minimal constraints. The company was allowed to join operations with, or hold shares in, other telephone companies. But the company was prohibited from operating a telegraph service, or holding shares in other companies that were *not involved in the provision of telephone service*. Moreover, Bell was required to return to Parliament each time it sought to increase its capitalization, beyond prescribed limits, through the issuance of stocks or bonds. Between its incorporation and 1968 Bell increased the capital of the company from $500,000 to $1,750,000. This required nine amendments to its charter. As Bell embarked on a corporate strategy to diversify its operations in the 1960s, it became imperative that the company find a way to remove the restrictive provisions of its charter.

In 1967 Bell petitioned Parliament to modify the Bell Canada Special Act in two respects. First, the company sought a change in the wording of its mandate, in order to make it broader so that it would better reflect the company's mission as a "telecommunications" company, and not just a "telephone" company. Second, Bell wanted the charter modified so that it could hold more than *shares* in other companies. For example, Bell wanted the right to hold *bonds*, *debentures*, and *other securities* in other companies, with the limitation that the other companies be engaged in the same type of business as Bell (Babe 1990, 184–185). Although Bell was able to realize these changes, the legislative process resulted in some unexpected additions to the charter.[23] The revised Bell charter stipulated that the company was to "act soley as a common carrier, and shall neither control the contents nor influence the meaning or purpose of the message emitted, transmitted or received" (Bell Canada Special Act, S.C. 1977-78, c.44 D.5.(3)). Moreover, the revised charter prohibited the company or its subsidiaries from holding a broadcast license or even applying for a broadcast license (D.5.(2)). Parliament had reiterated and reinforced the historical separation of content and carriage. Bell was given more freedom to diversify, but diversification that had as its objective the creation of content for transmission over Bell's facilities would not be tolerated. Robert E. Babe (1990) notes that Bell was able to circumvent the restrictions on ownership of non-telephone companies by making sure that the acquisitions were undertaken by Northern Electric or telephone subsidiaries controlled by Bell. By 1976 Bell had interests in seventy-four different companies around the world (186).

An important part of Bell's diversification strategy involved the internationalization of its operations. One result was the creation of Bell Canada International (BCI) as an independent subsidiary of Bell Canada. BCI was created to export Bell's expertise in the planning and operation of telephone networks. In the mid-1970s, BCI together with Dutch and Swedish partners won a very lucrative contract to plan and supervise the construction of a modern telephone infrastructure in Saudi Arabia. According to Jean-Guy Rens (1993b), Bell's portion of the contract was valued at $3 billion, spread over ten years of work. At one point more than one thousand Bell employees were working in Saudi Arabia (43). The contract was signed in 1978 and renewed in 1983. Through its subsidiary BCI, the contract produced revenues for the parent company Bell Canada. It was inevitable that Bell and the CRTC would clash on the treatment of the revenues.

From CRTC's perspective, the revenues derived from the contract, BCI's revenues, should have been included as part of the parent company's total revenues. This was the case since BCI's employees were largely employees of the parent company on loan to the international subsidiary. Moreover, at the request of the Saudi Arabian government, the contract had been signed with the partent company, Bell Canada. The inclusion of revenues from a profitable operation in Saudi Arabia would have the effect of lowering Bell Canada's revenue requirement for its Canadian operations, resulting in lower rates for Canadian subscribers. Moreover, such an approach was justified by the fact that Canadian ratepayers had, in effect, subsidized the training of the Bell employees working in Saudi Arabia.

From Bell's perspective, one simply had to reverse the scenario, and imagine what would happen should the contract produce a deficit instead of a profit. Would the CRTC then be prepared to tranfer liability to Bell's subscribers in the form of higher rates? From Bell's perspective the key issue was risk. If the contract failed to produce profits, Bell's shareholders would suffer lower share prices and dividends. Since Bell's shareholders assumed the risk, they should be the ones to benefit if the project was successful. From Bell's perspective the CRTC wanted to have its cake and eat it too. From the CRTC's perspective, monopoly revenues, in one form or another, had helped to put Bell in a competitive position to win the contract. Bell's ability to rely on expertise gained in the context of its Canadian monopoly was no doubt a factor in reducing the company's level of risk in the project. Ultimately, the CRTC backed away from its position, allowing the revenues to remain with BCI. But it imposed a 25 percent levy on the salaries of Bell employees working for BCI, to be charged against BCI in favor of the parent company (Rens 1993b, 45). More than anything else the BCI case reveals the difficulties involved in reconciling monopoly operations with corporate diversification.

In the spring of 1982, Bell Canada embarked on a new strategy, which it hoped would finally solve the problem of regulatory interference with corporate diversification. The solution was to reorganize the company in a way that would separate Bell's regulated common carrier business from its other unregulated business activities, thus enhancing Bell's autonomy as a corporation. It is interesting to note that this occurred at the juncture where AT&T in the United States was about to embark on its own reorganization/divestiture pursuant to a settlement of an antitrust suit launched by the U.S. Department of Justice in 1974. Bell Canada began by applying to the Department of Consumer and Corporate Affairs for a change in corporate status: it annulled its existing charter in order to become a corporation under the Canada Business Corporations Act (1985 and 1994). It then undertook a corporate reorganization that resulted in the creation of a new parent company, Bell Canada Enterprises (BCE), and brought the common carrier operations of Bell Canada under the BCE corporate umbrella.

The maneuver was part genius, part theatre of the absurd. Under the old structure, Bell Canada was the holding company for a number of subsidiaries. One of these was Tele-Direct, a publishing venture responsible, among other things, for printing Bell's directories and Yellow Pages. Because Bell could not invest directly in such a non-telephone operation, Tele-Direct was owned by a tiny local telephone company, the Capital Telephone Company. Capital Telephone operated in Maberly, Ontario, and had 121 subscribers (Babe 1990, 183). Bell controlled Tele-Direct indirectly through its ownership of Capital Telephone. This was permitted by law since Bell was allowed to hold shares in other telephone companies. The reorganization was realized first by releasing Tele-Direct from Capital Telephones. Tele-Direct then changed its name to Bell Canada Enterprises. A new company was formed, Tele-Direct (Canada) with responsibility for some of the former Tele-Direct's operations. This company became a subsidiary of the new holding company BCE. Bell Canada, the former holding company, also became a subsidiary of BCE. The telephone directory and Yellow Pages business of the former Tele-Direct were consolidated into a new corporation Tele-Direct (Publications), which became a subsidiary of Bell Canada, now a subsidiary of BCE. This was done because the CRTC had determined that directory and Yellow Pages were an integral part of the monopoly telephone business (Rens 1993b, 47). Babe has referred to this ironically as "birthing a parent" (1990, 190).

The reorganization was completed on April 28, 1983. Although the government requested that the CRTC assess the reorganization and make recommendations for legal safeguards to ensure that the new structure would not open the door to an abuse of the Bell Canada monopoly, the recommendations were not acted upon for four years.

On June 25, 1987, the Bell Canada Special Act of 1880, which had been amended numerous times, was repealed and replaced by the Bell Canada Act. The new act, which prevailed over the Canada Business Corporations Act, introduced a number of constraints on the business operations of Bell Canada. The act continued the prohibition on Bell Canada's right to hold a broadcasting license (article 8). However, given the new structure, this control was rather weak since there was nothing to prohibit the parent company, BCE, from holding a broadcasting license through a separate subsidiary. Article 10 stipulated that Bell Canada could not hold shares in any company that is either engaged in the manufacture of products for sale to the company, or involved in research and development of such products. However, Bell Canada could retain its holdings in its subsidiary, Bell Northern Research. Article 13 contained more teeth. It empowered the CRTC to order Bell to undertake telecommunications operations previously carried out by an unregulated subsidiary, if the commission determines that the subsidiary was not subject to an adequate degree of competition (13(1)). This was designed to discourage excessive and unfair competition designed to stifle competition. The CRTC was also empowered to order Bell to divest itself of any activity that it deems to be competitive (13(2)). These provisions were designed to discourage Bell Canada from spinning off profitable monopoly operations and to ensure that Bell would not mix its competitive and its monopoly businesses. Finally, the act stipulated that the parent corporation, BCE, must provide the CRTC with any information it requires in order to exercise its responsibilities under Canadian law.

7.8 THE PRIVATIZATION OF TELEGLOBE

On October 30, 1984, the government announced its intention to privatize Teleglobe, Canada's overseas telecommunications carrier. Teleglobe's predecessor, the Canadian Overseas Telecommunications Corporation (COTC), was created in 1949 through an act of Parliament: the Canadian Overseas Telecommunication Corporation Act (1949). The act was a direct consequence of the Commonwealth Telegraphs Agreement signed in May 1948. Under the terms of the accord, the member countries of the British Commonwealth agreed that each country would designate a government department or public corporation to run its international telecommunications oper-ations. This corporation would then be charged with representing its govern-ment at meetings of the Commonwealth Telecommunications Board. The COTC was established as a publicly owned crown corporation. Its first man-date, stipulated in the act, was to purchase the Canadian assets of the Cana-dian Marconi Company and of Cable Wireless Limited of London.

The COTC was created as a *carrier's carrier* with a mandate to provide telecommunications service in the overseas market. As a carrier's carrier, the corporation was mandated to provide telecommunications capacity (chan-nels) to Canada's first-line telecommunications providers. The corporation provided, and continues to provide, international telecommunications links to Canada's domestic telecommunications providers via transoceanic under-water cables and satellite circuits. Rather than providing service directly to individual subscribers and businesses, COTC leased international circuits to domestic carriers such as Bell Canada and CNCP. The COTC's monopoly ex-tended to all overseas communication, but it did not include continental communications between Canada and the United States. Telecommunica-tions links between Canada and the United States were negotiated directly between Canadian and U.S. domestic carriers.

In addition to its responsibilities as Canada's international carrier, the COTC was charged with representing Canada internationally at Intelsat and at International Maritime Satellite Organization (Inmarsat). In 1975 the COTC changed its name to Teleglobe Canada.

The idea of privatizing Teleglobe had been launched as early as 1979 by Jean-Claude Delorme, Teleglobe's president (Rens 1993b, 515–516). When the Conservative Party, led by Brian Mulroney, took power in September 1984, they were receptive to the idea of privatization. It warrants mention that Teleglobe was not a typical crown corporation: it was a profitable com-pany. For example, in 1986 Teleglobe produced a profit of Can $63 million on revenues of Can $813.5 million (Gillick 1987, 213). There was no reason to believe that it would not continue to be profitable in the future. Although the government was eager to sell, it did not want the privatized company to be controlled by Canada's existing domestic carriers and, in particular, by Bell Canada. In order to prevent this, the government announced that the total number of shares that could be held by common carriers would be limited to 33 percent. Foreign *telecommunications carriers* were not permited to hold shares in the company, and foreign ownership in general would be limited to 20 percent of voting shares as was the case with Canada's other common car-riers.[24] After an intial offering failed, Teleglobe was put on the market a sec-ond time; this resulted in its sale to Memotec Data. The privatization was accomplished through the passage into law of the Teleglobe Canada Reorga-nization and Divestiture Act on April 1, 1987. The actual sale of Teleglobe to Memotec took place on April 4.

Memotec was an unlikely buyer to say the least. With its headquarters in Montreal, Memotec was an international company providing data communications and data processing services to businesses. Among its products were multiplexers used by telecommunications operators for packet-switching. The book value of Memotec's assets at the time of the sale was estimated at Can $83.6 million (Gillick 1987, 214). How could a company valued at $84 million purchase a company valued at $524.9 million? The answer was a typical 1980s leveraged-buyout. Memotec financed its purchase with a bridging loan of $225 million, the sale of convertible debentures for $125 million, and a future offering of common shares. Teleglobe's outstanding debt to the government of $143 million would be refinanced by the National Bank of Canada. The government rejected bids by Power Corp of Canada, and from a consortium involving the Caisse de dépôt et placement du Québec, Spar Aerospace, and a number of Canadian telephone companies.

It was unrealistic to expect Memotec to be able to carry a debt of such magnitude for very long. In May of the same year BCE was back in the picture after a friendly purchase of 30 percent of Memotec's capital (Rens 1993b, 517). It was not long after the privatization that problems began to emerge with Memotec's management of Teleglobe. Both the CRTC and BCE were dissatisfied with Memotec's management of this national telecommunications asset. Rens (1993b) notes that Memotec used Teleglobe reserves to finance the purchase of three U.S. computer services companies valued at $130 million. The CRTC objected to the manner in which Teleglobe's accounts were merged with Memotec's centralised accounting system (517). Perhaps most important, BCE was unhappy with William M. McKenzie's leadership of Memotec. Under McKenzie, Memotec had begun to develop services that systematically bypassed Canada's domestic carriers. A struggle for control of Teleglobe ensued. In January 1992, BCE, with the assistance of the Caisse de dépôts et de placement du Québec, succeeded in ousting McKenzie and replacing him with Charles Sirois. In 1997, BCE controled 24.3 percent of Teleglobe shares.

Under the terms of the Teleglobe Canada Reorganization and Divestiture Act, it was foreseen that Teleglobe would retain the exclusive right to carry overseas traffic originating or terminating in Canada until at least March 31, 1992.[25] In 1992 the minister of communications announced that Teleglobe's monopoly would be extended to April 1997. On the same occasion it was announced that the government would undertake a review of Teleglobe's mandate in 1995. Key among the questions considered in the 1995 review would be Teleglobe's monopoly. Specifically, the review would consider whether facilities-based competition should be allowed in the market for overseas telecommunications services. Other issues to be considered included the following: (1) whether Teleglobe should continue to represent Canada as the signatory to Intelsat and Inmarsat agreements; and (2) whether the ownership restrictions that limited Canadian carrier ownership to 33 percent and foreign ownership to 20 percent should be removed or varied (Industry Canada 1995b).

Since its privatizatization, Teleglobe has moved aggressively into the international market for telecommunications services. In 1996 Teleglobe's U.S. subsidiary, Teleglobe USA, received a license from the FCC to operate as a facilities-based carrier in the United States. On June 12, 1997, Teleglobe was listed for the first time on the New York Stock Exchange. The company has not resisted plans to end its monopoly. Rather, it has urged the government to negotiate aggreements with other countries based on the principle of reciprocity. Teleglobe's access to foreign markets can only be gained at the price of liberalizing the Canadian overseas markets. Because Teleglobe's plans for the future lie in international expansion, it has supported reciprocal international liberalization.

On February 5, 1997, the World Trade Organization reached an agreement on basic telecommunications. The Canadian government, according to a government issued press release, welcomed and actively promoted the agreement. Under the terms of the deal, Canada pledged to reverse key provisions of the Teleglobe Canada Reorganization and Divestiture Act that restricted ownership of Teleglobe and protected the company's monopoly. Also, Canada was prepared to open the domestic satellite market to facilities competition: Telesat's monopoly would cease to exist on March 1, 2000. According to the press release, Canada would take the following steps to further liberalize its telecommunications industry:

- End Teleglobe Canada's monopoly on overseas traffic on October 1, 1998
- End Teleglobe's special ownership restrictions that prohibit investment by foreign telecommunications carriers and limit the investment by Stentor
- Allow 100 percent on foreign ownership and control in the resale sector
- Allow 100 percent on foreign ownership and control of international submarine cable landings in Canada, as of October 1, 1998
- Remove all restrictions on the use of foreign owned and controlled global mobile satellites providing services to Canadians as of October 1, 1998
- End the Telesat monopoly on fixed satellite services effective March 1, 2000
- Allow the use of any foreign satellite to provide services (other than Direct to Home/Direct Broadcast Satellite) to Canadians, as of March 1, 2000
- Remove traffic routing rules for all international services and all satellite services by March 1, 2000
- Maintain its open, competitive market and existing transparent regulatory regime (Industry Canada 1997)

It is extremely important to note that the agreement covered "basic telecommunications services," and did not include broadcasting services. For example, direct-to-home satellite services were not covered by the agreement. As a result of the agreement, Canadian companies such as Teleglobe, Telestat, Bell Canada, and the other domestic carriers, will have greater access to overseas markets, both in terms of opportunities to invest in foreign carriers and in terms of opportunities to offer services in foreign markets.

7.9 THE TELECOMMUNICATIONS ACT OF 1993

On June 23, 1993, Canada's first *consolidated* telecommunications statute received royal assent. The Telecomunications Act represented a major housekeeping of federal telecommunications legislation. The act abolished the National Telecommunications Powers and Procedures Act, the Telegraphs Act, and displaced the provisions of the Railway Act dealing with telegraph and telephone communications. The new act also amended the following legislation: the Bell Canada Act, the Telesat Canada Act, the CRTC Act, and the Teleglobe Canada Act. The most significant accomplishment of the Telecommunications Act was the introduction, for the first time, of a statement of telecommunications policy objectives. Although the Broadcasting Act contained a set of policy objectives for the broadcasting sector, there was no equivalent statement of principles for the telecommunciations industry.

For the most part, the Telecommunications Act simply continued the provisions of earlier statutes (primarily the National Telecommunications Powers and Procedures Act and the Railway Act). However, the new act broke ground in seven areas:

1. the aforementioned statement of policy objectives (section 7);
2. its stipulation that carriers must be Canadian owned and controlled (section 16);
3. its clarification of the respective roles and responsibilities of the CRTC, the minister, and the Governor in Council (sections 8–15);
4. the act's provision for the exemption of classes of carriers from the act (section 9);
5. its resolution of the problem of federal jurisdiction to regulate provincial crown corporations;
6. its requirement that the minister consult with the provinces with respect to certain provision of the act; and (most importantly)
7. the act's inclusion of the power to forbear from regulating.

Our review of the act is not meant to be comprehensive. Rather we will emphasize the provisions of the act that represent substantive change with respect to previous legislation.

Section 7 of the Telecommunications Act enumerates, for the first time, a set of objectives for Canadian telecommunications policy. Section 7 states:

It is hereby affirmed that telecommunications performs an essential role in the maintenance of Canada's identity and sovereignty and that the Canadian telecommunications policy has as its objectives:
  (a) to facilitate the orderly development throughout Canada of a telecommunications system that serves to safeguard, enrich and strengthen the social and economic fabric of Canada and its regions;

(b) to render reliable and affordable telecommunications services of high quality accessible to Canadians in both urban and rural areas in all regions of Canada;
(c) to enhance the efficiency and competitiveness, at the national and international levels, of Canadian telecommunications;
(d) to promote the ownership and control of Canadian carriers by Canadians;
(e) to promote the use of Canadian transmission facilities for telecommunications within Canada and between Canada and points outside Canada;
(f) to foster increased reliance on market forces for the provision of telecommunications services and to ensure that regulation, where required, is efficient and effective;
(g) to stimulate research and development in Canada in the field of telecommunications and to encourage innovation in the provision of telecommunications services;
(h) to respond to the economic and social requirements of users of telecommunications services; and
(i) to contribute to the protection of the privacy of persons.

From the perspective of our study of telecommunications liberalization, the sixth objective (f) is the most significant. The inclusion of this objective makes it clear, for the first time, that the CRTC has a statutory mandate to promote competition. Also of import are the objectives promoting Canadian ownership of facilities (d) and the use of Canadian facilities for telecommunications (e).

More substance is given to the objective relating to Canadian ownership of facilities in section 16 of the act. The act states: "A Canadian carrier is eligible to operate as a telecommunications common carrier if it is a Canadian-owned and controlled corporation incorporated or continued under the laws of Canada or a province." In order to meet the requirement of being Canadian owned and controlled, the following conditions must be met: 80 percent of the company's board of directors must be individual Canadians; Canadians must own not less than 80 percent of the voting shares of the company; and the company must not be otherwise controlled by persons who are not Canadian (Section 16, 3.(a)(b)(c)). However, foreign ownership of a *holding company* may reach 33⅓ percent of voting shares (Darling 1996).

Under the Telecommunications Act the Governor in Council retains its power to vary, rescind, or return a decision to the commission either through its own initiative, or in response to a petition brought by another party (section 12.1). Section 8 grants the Governor in Council new powers to issue to the commission "directions of general application on broad policy matters with respect to the telecommunications policy objectives." This means that the federal cabinet has the power to give further interpretation to the policy objectives stated in the act, thus providing additional guidance to the CRTC in its deliberations. The act also grants new powers to the CRTC. Section 9 confers the power to exempt "any *class of Canadian carriers* from the application of the [A]ct" (emphasis added). For example, the commission could exempt the class of radio common carriers (cellular telephone companies) from the obligation to file tariffs, should it determine that the sector was competitive. As was the case under the National Telecommunications Powers and Procedures Act, a decision of the commission may be appealed to the courts on a point of law or jurisdiction (64.1).

Section 3 of the act resolves once and for all the question of federal jurisdiction to regulate common carriers that are provincial crown corporations, such as SaskTel and the Manitoba Telephone System. The act states, "This [A]ct is binding on Her Majesty in right of Canada or a province." Recall that the Supreme Court's AGT decision resolved the question of crown immunity when it determined that crown immunity could be dissolved through an explicit statement of such an intention in federal legislation. On the contentious question of federal/provincial consultation, the act requires the minister to notify a designated minister of each province before making a recommendation to the Governor in Council concerning the following: (1) directions to the commission concerning broad policy (section 8); (2) the exemption of classes of carriers (section 9); (3) varying, rescinding, or returning a decision to the commission (section 12); and (4) establishing technical standards (section 15). The act does not include a right of provincial participation in the deliberations of the CRTC. Nor are the results of federal/provincial consultation binding upon the minister. However, a revision to the Broadcasting Act in 1991 prescribed that regional members of the commission would be appointed from the following regions of Canada: Atlantic Provinces, Quebec, Ontario, Manitoba and Saskatchewan, Alberta, and British Columbia.

The reader will recall from chapter two, section four, that the regulation of common carrier rates, the obligation of common carriers to file tarrifs before the commission, and the stipulation that rates be reasonable and non-discriminatory were contained in the Railway Act. These provisions were carried over to the Telecommunications Act almost verbatim. Section 24 states that all services provided by a common carrier are subject to any conditions imposed by the commission or approved by the commission in a tariff. Section 25 requires all Canadian carriers to file tariffs with the commission for the services they provide. The tariffs must specify the maximum or minimum rate (or both) to be charged for the service. Section 27 requires that rates be just and reasonable and prohibits a carrier from unjustly discriminating or giving undue preference toward any person, including itself. Under section 29, agreements between common carriers concerning the interchange of traffic, the managment of facilities, and the apportioning of rates or revenues between carriers are subject to commission approval.

Sections 24, 25, 27, and 29 of the Telecommunications Act essentially reiterate requirements that had already been entrenched in the Railway Act. But the new act gives the commission the power to refrain from applying them. Section 34 (1) gives the commission the power *not to regulate* in respect of sections 24,

25, 27 and 29. The commission now has the freedom not to regulate a service or class of services where it determines that the service is subject to competition. In regulatory discourse the power not to regulate is also referred to as forbearance. To the extent that one of the original objectives of the pro-competitive movement was to reduce or eliminate regulation altogether, the power to forbear from regulating must be considered an important step towards deregulation. It remains to be determined how often and with what results the commission will be able to exercise forbearance.

7.10 CONCLUSION

In this chapter we have reviewed the long process through which the Canadian government, including the DOC and the CRTC, liberalized important sectors of the Canadian telecommunications industry. Although cultural, national sovereignty, and national unity themes permeated official government discourse on the future of telecommunications, we have seen that by the mid-1970s Canada had begun to chart a course designed to meet the changing conditions and requirements of large users of telecommunications. When this occurred, the focus of attention shifted to the needs and demands of both producers and consumers of telecommunications services. These included the following: equipment manufacturers, potential competitors to network operators, Canadian business users of telecommunications, and, to a lesser degree, Canadian residential consumers. The focus of social policy for telecommunications shifted from the national policy triad of cultural, sovereignty, and national unity objectives, to objectives that were closer to the actual activities and preoccupations of the telecommunications industry: universality, access, affordability, fairness, efficiency, competitiveness, the role of regulation, and, most important, industry structure.

The shift from monopoly to competition in the terminal equipment market took fifteen years to complete. It began in 1967 when Dr. H. S. Gellman petitioned Parliament to revise Bell's charter in order to require the company to establish rules for the attachment of foreign equipment to its network. Telecom Decision CRTC 82-14, *Attachment of Subscriber-Provided Terminal Equipment*, effectively opened the terminal equipment market to full competition in 1982. Change was slow because Bell Canada fiercely resisted it. A key to the decision to liberalize the terminal equipment market was the CRTC's conclusion that competition in this market would not be harmful to Bell Canada's overall profitability and, therefore, would not threaten universality.

In a similar fashion, the liberalization of the long distance market did not occur overnight. It took over fifteen years to accomplish, again as a result of Bell Canada's attempts to preserve its monopoly. In 1976 the CRTC requested interconnection with Bell Canada so that it could offer a long distance data and private-line service. The application was approved in 1979 (TD CRTC 79-11). However, it was not until 1992 that the CRTC approved an application to provide a public long distance voice service (TD CRTC 92-12). As was the case with terminal equipment, the CRTC determined that competition in MTS–WATS would not present a significant threat to universality. However, as we will see in chapter eight, long distance competition would lead to rate rebalancing, which would produce rate increases for local telephone service. Perhaps the most significant feature of the long distance decision was the ostensible shift from a policy designed to reduce the price of MTS–WATS to one emphasizing choice and innovation. But the real shift in policy concerned the heightened responsiveness to the demands of Canada's corporate users to the detriment of the class of residential users, and in particular to the detriment of low-income residential users.

The liberalization of the terminal equipment and long distance markets was largely orchestrated by the CRTC. The commission would play a more peripheral role in the Bell Canada reorganization, the privatization of Teleglobe, and the passage of the Telecommunications Act of 1993.

Finally, it warrants mention that the CRTC was not given a totally free hand to liberalize. Efforts to promote competition were occasionally stifled by government. When the CRTC, shortly after it was given authority to regulate telecommunications, attempted to introduce competition to the domestic satellite market, its efforts were thwarted by the federal cabinet with the support of the members of Telecom Canada.

What has been the impact of U.S. deregulation on the Canadian telecommunications sector? We have seen that the impact of U.S. deregulation was neither immediate nor direct. The deregulation of the Canadian terminal equipment market occurred in 1982, approximately five years after the U.S. market had been liberalized. Long distance competition was introduced in 1992, fully fifteen years after the *Execunet* decision in the United States. In general the Canadian approach has been less ideological, and, particularly during the early period of liberalization, more sceptical about the benefits of competition.

In large measure, the more prudent Canadian approach was a consequence of the demonstration effect of the U.S. experience that revealed the shortcomings of aggressive deregulation. There were clearly winners and losers in the United States: residential users, who rely primarily on local service, saw their rates increase substantially; corporate users, who rely more heavily on long distance service, were the principal beneficiaries (Crandall 1990, 65). Also, the fact that U.S. rate rebalancing led many states to implement lifeline services made it more difficult for the proponents of competition in Canada to argue that liberalization reduces regulation and has no significant impact on universality. In order to ensure that economically disadvantaged families would still be able to afford local telephone service following local rate increases due to rate rebalancing, forty-nine states had implemented a lifeline telephone subsidy by 1991 (Mosco 1991). Individuals who passed a needs test administered by the government were able to qualify for this subsidy.

The Canadian response to U.S. deregulation was also tempered by a characteristic Canadian reluctance to place too much faith in the marketplace. The Canadian political scientist Richard Schultz described this disposition in the following terms:

> At a general level, Canadians and their governments are less willing to rely on, or trust, the competitive marketplace to achieve particular goals. There are both functional and philosophical reasons for this. In part it reflects a belief that the Canadian marketplace is neither large enough, nor strong enough, particularly vis-à-vis that of the United States, to rely on competition. In part our more limited reliance on markets and our stronger roles for governments reflect, if not a rejection, at least a tempering of the values or goals that market place competition represents. (1984, 62)

In a similar vein, John Meisel, a former chairman of the CRTC, argued that U.S. deregulation did not apply to the Canadian context. Meisel stated:

> The US is now experiencing a wave of deregulatory fervour which, although less all-embracing than is sometimes claimed is certainly serious. It has led the Federal Communications Commission and other regulatory agencies to relax or completely abandon some of their former customary activities. While it is too soon to speak of the results of this policy, it is plain that it is not applicable in a Canadian context. In this country, we have always tempered the desire for economic prosperity with a broad, public concern for the achievement of political and social ends. These, over the long haul, have held firm and our economic successes have been traditionally marshalled in their service. This fact is responsible in a very large measure for our survival as a country over a period in which the great majority of nation states have faltered and failed. (Canadian Radio-television and Telecommunications Commission 1982, ix)

Finally, as already discussed, regulatory fragmentation combined with provincial opposition to competition, slowed the pace of telecommunications liberalization in Canada.

NOTES

1. In 1977 and again in 1978 the government introduced legislation designed to consolidate federal telecommunications legislation. This legislation would have replaced the Broadcasting Act, the Telegraphs Act, the Canadian Radio-television and Telecommunications Act, the Radio Act, and the sections of the Railway Act and the National Transportation Act that applied to telecoms. These changes were proposed in Bill C-43 (first reading 22 March 1977), Bill C-24 (26 January 1978), and Bill C-16 (9 November 1978). None of these bills passed first reading.

2. The term "rate rebalancing" refers to a new regime for the pricing of telephone service based on the elimination of the subsidy from long distance to local service.

Under rate rebalancing, rates for long distance would be set closer to costs with the result that the price of long distance would be reduced. Similarly, rates for local service would rise to reflect their true costs.

3. It should be noted that opposition to competition in the Prairie Provinces was not monolithic. Alberta was historically more open to competition than the provinces of Manitoba and Saskatchewan, although it was equally concerned with the lack of political controls on the CRTC.

4. The term "programming services" refers to services, such as traditional cable TV, where the content is offered as part of a scheduled sequence of programs. These are typically video services. An example of a non-programming service would be access to a computer-based service where the consumer determines what he or she will access and when he or she will do so. This type of content is usually stored as lines of text or computer graphics.

5. During the course of the litigation, TCTS would change its name to Telecom Canada (1983). On January 29, 1992, Telecom Canada underwent another reorganization and became known as Stentor Canadian Network Management. These reorganizations and name changes did not affect membership in the consortium that remained essentially the same.

6. Also known as the Constitution Act of 1867, the act was replaced by the Constitution Act of 1982.

7. DCF Systems was a small telecommunications consulting firm operating in Ontario (Rens 1993b, 295).

8. It was not until April 1976 that responsibility for telecommunications was transferred from CTC to CRTC.

9. New title, Competition Act. R.S., 1985, c. C-34, s. 1; R.S., 1985, c. 19 (2nd Supp.), s. 19.

10. The company's name was changed to Northern Telecom on March 1, 1976. Today it is known as Nortel.

11. It is important to make a distinction between the RTPC and the director. The RTPC oversees the director, who reports to and makes recommendations to the commission.

12. See chapter two for more details on the history of collaboration between Canadian National and Canadian Pacific, which led to the formation of CNCP Telecommunications in 1980.

13. MTS is telephone jargon for a direct dial (i.e., without the intervention of an operator) long distance toll call. Long distance calls are *toll* calls because they are billed on the basis of distance and time.

14. On November 24, 1981, a similar decision was made with respect to interconnection with B.C.Tel (*CNCP Telecommunications: Interconnection with the British Columbia Telephone Company*, TD CRTC 81-24). As a result of its success in obtaining interconnection with Bell and B.C.Tel, CNCP then applied to the CRTC for similar interconnection arrangements with AGT. It was this application that set in motion the appeal process in the courts which resulted in the Supreme Court decision that settled the question of federal/provincial jurisdiction over telecommunications.

15. The reader is probably most familiar with WATS as a "1-800" service, where a long distance call is made at the expense of the receiver of the call. WATS is a *public* telephone service. A company/organization that provides WATS as a service to its customers is billed at a discount rate based on the number of minutes of use during a month.

16. BCRL's application was also considered in the same proceeding. Lightel was a unit of Call-Net Telecommunications, Canada's largest telecommunications reseller.

17. Because Bell's switching facilities had not yet been upgraded to provide "equal access," the first Unitel subscribers had to dial extra digits. This would change on July 1, 1994, when Unitel subscribers were able to access long distance with the same ease as Bell and B.C.Tel subscribers.

18. Canadian Overseas Telecommunications Corporation was created by an act of Parliament, the Canadian Overseas Telecommunications Corporation Act of 1948. In 1975 the corporation changed its name to Teleglobe Canada.

19. It is important to note that the COTC operated satellites and undersea cables. Unlike Telesat, the COTC *was not restricted to using satellite technology*. Its monopoly was not defined by a specific technology, rather it was defined by a particular market sector: overseas telecommunications.

20. Telesat was not established as a crown corporation such as the CBC or Canadian National Railways. Crown corporations, which are effectively owned by a province or the federal government, are not managed as a department of government.

21. 14 GHz refers to the band used for receiving signals from the satellite. 12 GHz refers to the band employed to transmit signals to the satellite. The first *Anik* satellites operated at 6/4 GHz.

22. Between 1882 and 1968 the Bell charter was amended no less than twelve times (Department of Communications 1971a, 31). The version we have described here reflects the status of the charter in the mid-1960s when Bell began to lobby for greater flexibility in the law.

23. The revised charter stated: "the Company has the power to transmit, emit or receive and to provide services and facilities for the transmission, emission or reception of signs, signals, writing, images or sounds or intelligence of any nature by wire, radio, visual or other electromagnetic systems and in connection therewith to build, establish, maintain and operate, in Canada or elsewhere . . . all services and facilities expedient or useful for such purposes" (D.5.(1)).

24. The limitation of 20 percent of voting shares by foreign interests was enshrined in the Telecommunications Act of 1993 and would thereafter apply to all Canadian carriers.

25. This did not apply to international traffic between Canada and the United States. Teleglobe could participate in this market but without a legislated monopoly.

8

From Enhanced Services to Local Network Competition: The Canadian Approach

8.1 INTRODUCTION

The Canadian Radio-television and Telecommunications Commission (CRTC) approached the problem of enhanced services in two stages. It began by approving common carrier requests to engage in market trials[1] for enhanced services, but only on a temporary basis. These market trials were approved on a case-by-case basis, and were viewed by the commission as interim decisions. The second stage involved a comprehensive public hearing that would deal with the regulatory treatment of enhanced services. The public proceeding resulted in a decision issued on July 12, 1984, entitled simply, *Enhanced Services* (TD CRTC 84-18). As we will see in this chapter, the Canadian approach to enhanced services borrowed from the American model in one important respect: it was based on the distinction between basic and enhanced services outlined in the *Computer II* decision. However, it also diverged from the American model when it did not require Bell Canada to offer enhanced services through a structurally separate affiliate.

In the next stage of Canadian policy making, which followed the liberalization of long distance markets, the CRTC addressed the problem of cable/telephone convergence. In doing so, the commission went beyond the issue of common carrier diversification in order to consider enhanced services and to decide whether common carriers should be allowed to enter markets for TV programming services under the Broadcasting Act. Conversely, the commission would consider whether the cable industry should be allowed to participate in markets for local telephone service. The result was a new pro-competitive regulatory framework that linked telephone industry diversification into new markets for broadband services to the removal of the regulatory controls that had protected the telephone companies from competition in local networks. As a result of the policy, the last pillar of the old telephone monopoly, the local network bottleneck, would be eliminated.

8.2 THE CRTC AUTHORIZES COMMON CARRIER PROVISION OF ENHANCED SERVICE ON A TEMPORARY BASIS

In September 1980 Bell Canada, on behalf of the Trans Canada Telephone System (known as Telecom Canada after 1983), submitted tariffs to the CRTC for an enhanced service known as Voice Message Service (VMS). This was the commission's first opportunity to formulate policy on the new hybrid services that combined communications and data processing. VMS was an enhanced *voice* service that allowed subscribers to call a specific telephone number and record a message; the message could then be delivered using the telephone network to destinations (telephone numbers) identified by the subscriber. In a decision rendered on May 25, 1981, entitled *Bell Canada—Voice Messaging Service* (TD CRTC 81-10), the CRTC granted interim permission to provide the service as part of a market trial. In early 1983 Bell withdrew the service, citing non-profitability as the cause (Department of Communications 1983b, 10). In September 1981 a similar service was introduced by a division of Shell Canada Limited. The service was known as Voice Message Exchange (VMX). Because the service was not run by a common carrier, CRTC approval was not required (10).

The first application to provide an enhanced *data* service was submitted to the CRTC by Bell Canada in 1981. Known as Envoy 100, the service was a text-based messaging service, an early form of electronic mail (e-mail). It was approved on an interim basis by the CRTC on November 4, 1981, *Bell Canada—Envoy 100 Service* (TD CRTC 81-22). On February 10, 1984, while the general proceeding on enhanced services was underway, the CRTC approved an application for an eighteen-month market trial by Bell Canada of a service known as iNet 2000, *Bell Market Trial of iNet 2000* (Telecom Order CRTC 84-57). iNet 2000 was a *gateway service* that provided access to on-line databases in the United States and Canada. The service was designed for the business market. The iNet 2000 service provided a directory of these databases, which included a schedule of fees to access them. It could also act as a billing agent for the database providers. iNet 2000 included a messaging service (e-mail) so that its subscribers could communicate with subscribers to Envoy 100 service.

Bell Canada was not the only common carrier to seek regulatory approval to offer an enhanced service. On March 5, 1982, the commission approved a Canadian National and Canadian Pacific (CNCP) Telecommunications tariff application to initiate a market trial for a service known as Infotex, *CNCP Market Trial of Infotex Service* (Telecom Order CRTC 82-116). Infotex was a service that permitted data communication between word processors via CNCP's Infoswitch data network (Department of Communications 1983b, 11). CNCP also received approval to offer an information service known as Telenews, which was available to CNCP Telex users. By dialing a specific number, subscribers could receive news reports from a subsidiary of Canadian Press on their Telex data terminals. The service was approved by the CRTC on June 18, 1981, in *CNCP Telecommunications—TELENEWS Service* (TD CRTC 81-12).

When the CRTC authorized the first enhanced services, it did so without requiring the common carriers to operate them through separate subsidiaries. This was possible for two reasons. First, these were market trials; the common carriers would have to return to the commission for tariff approval before a true commercial service could be offered. Second, the commission was in the process of creating a general policy for enhanced services. Once the results of this proceeding were known, they could be applied to the services that were being offered on a trial basis.

8.3 THE CRTC INITIATES A GENERAL PROCEEDING ON ENHANCED SERVICES

The second stage began on November 15, 1983, when the CRTC announced that it would initiate a public proceeding entitled, *Enhanced Services* (Telecom Public Notice CRTC 1983-72). The issues to be considered in the proceeding included the following: the definition of enhanced services, the regulatory treatment of enhanced services provided by common carriers, the regulatory treatment of enhanced services provided by non-common carriers, resale, sharing, and interconnection for the purpose of providing enhanced services.

The commission received a large number of submissions from a diverse group of interested parties. In addition to the common carriers (Bell Canada, the other Telecom Canada members, CNCP, and Telesat) these parties included the provincial governments; the Canadian Newspaper Publishers Association; the Canadian Bankers' Association; the Telephone Answering Association of Canada; the Director of Investigation and Research, Combines and Investigations Act; the Consumers' Association of Canada; and the Canadian Manufacturers' Association, to name a few.

The reader will recall that the Federal Communications Commission (FCC) issued its *Computer II* decision in 1980. When the CRTC issued its enhanced services decision on July 12, 1984, it essentially copied the U.S. framework based on two service categories: basic and enhanced (TD CRTC 84-18). Although Bell Canada and the other common carriers opposed this approach, most of the other participants in the proceeding endorsed it. The definitions of basic and enhanced services that were adopted by the CRTC were essentially the same as their U.S. counterparts. Basic service was defined as "one that is limited to the offering of transmission capacity for the movement of information." An enhanced service was defined as "any offering over the telecommunications network which is more than a basic service" (12 and 14).

The definitional approach, based on the basic/enhanced dichotomy, left a number of activities that were essential to basic transmission, but which entailed some form of processing of information, in a regulatory grey zone. Techniques such as error control, compression, and protocol conversion were by this time routinely applied to voice communications in order to lower the costs and improve the efficiency of the public network. It was becoming more and more difficult to speak of "pure" transmission. But the commission did not wish to classify services as enhanced simply because they were subject to some form of processing during transmission. The commission attempted to avoid confusion by stipulating that the use of these techniques would not necessarily alter the status of a basic service. The commission stated the following: "Use internal to the service provider's facility of companding techniques, bandwidth compression techniques, circuit switching, message or packet switching, error control or other techniques that facilitate economical, reliable movement of information does not alter the nature of the basic service. Similarly, internal speed, code and protocol conversion that is not manifested in the outputs of the service does not alter the nature of the basic service" (TD CRTC 84-18, 13).

In other words, the use of these techniques would not automatically result in a determination that a service was enhanced. According to the commission, it was not the use of these techniques that determined a service's classification, but rather, the uses *that these techniques were put to.* If these processing applications were used to provide subscribers with "additional, different, or restructured information," or if the service involved subscriber interaction with the information (such as in a database), then a service would be classified as an enhanced one. On the other hand, in a basic service, "a service provider essentially offers a pure transmission capability over a communications path that is virtually transparent in terms of its interaction with subscriber supplied information" (TD CRTC 84-18, 14).

The Canadian definitions proved to be just as difficult to apply as their U.S. counterparts. This became apparent when the commission began to actually classify services in Telecom Decision CRTC 85-17, *Identification of Enhanced Services.* In that proceeding, groups representing diverse Canadian business interests[2] challenged Bell Canada's and British Columbia Telephone's (B.C.Tel) interpretation of the classification scheme, arguing that a number of services identified as basic by the common carriers were actually enhanced offerings. The commission was obliged to evaluate what it described as a "large number of services" on case-by-case basis (4). It would be inaccurate to state that the framework was totally unworkable, since the commission found it to be a useful basis for deciding whether specific services should be classified as basic or enhanced. There was probably less dissatisfaction with the definitional approach in Canada than in the United States. However, this was not due to different attitudes towards the definitional approach itself. Rather, it was a consequence of the CRTC's decision not to require common carriers to offer enhanced services through structurally separate affiliates.

In the United States, the FCC had used the basic/enhanced distinction to identify the services that common carriers would have to provide on a separate basis. In the case of AT&T, "maximum" structural separation was required. In the case of GTE, and the other common carriers, a lesser degree of separation was in order, one that prescribed that they simply keep a separate book of accounts. In the enhanced services decision, the CRTC made it clear that it was not inclined to create a separate regulatory framework just for Bell Canada. It preferred to apply the same framework to the regulation of Bell's enhanced services as it would to the other common carriers. Moreover, the commission decided that it would be appropriate to deal with the issue of structural separation in a subsequent proceeding. Pending that proceeding, Bell and the other common carriers would be allowed to provide enhanced services in-house, subject to reporting (accounting) requirements designed to prevent cross-subsidies.

Although the CRTC announced that it would consider structural separation at a later date, it never followed through with its statement. As a result, the common carriers under CRTC jurisdiction were allowed to provide enhanced services through the regulated companies. However, regulatory approval was required for each new enhanced service. Approval was contingent upon the carrier demonstrating that the enhanced service was financially viable; accounting controls were required to ensure that the service would not benefit from subsidies drawn from the company's monopoly business. Basic service components would be provided to the enhanced service at tariffed rates, and would *not* take into account any economies that existed due to in-house operations. This was designed to ensure fair competition.

It warrants noting that Bell Canada and the other common carriers were not content with a framework that allowed them to offer enhanced services as part of their regulated operations. They also wanted the freedom to spin off these operations into separate (unregulated) subsidiaries in the event that they should one day become profitable. This was the regulatory equivalent of "having your cake, and eating it too." Essentially, the commission acquiesced to this demand when it decided "not to impose, at this time, any restrictions on transfers to a separate affiliate of enhanced services provided by carriers on an in-house basis" (TD CRTC 84-18, 54). Instead, the commission announced that it would scrutinize the transfer of assets to separate affiliates in order to ensure that monopoly revenues had not subsidized the development of these services.

Enhanced services combine communications (transmission) with some form of data processing. But the definition of data processing did not extend to applications that involved "editorial control over content" (TD CRTC 84-18, 32). The Bell Canada Special Act (1977–1978) endorsed the principle of separation of content and carriage. Subsection 5(3) of the act stipulated the following: "The Company shall, in the exercise of its power under subsection (1), act solely as a common carrier, and *shall neither control the contents nor influence the meaning or purpose* of the message emitted, transmitted or received as aforesaid" (emphasis added). The CRTC interpreted this statement to mean that Bell Canada could provide enhanced services, but it could not engage in electronic publishing; nor would the company be authorized to create or distribute its own databases (35–36). Moreover, the commission determined that this principle should be extended to the other federally regulated common carriers, that is, B.C.Tel, CNCP, and Telesat.

The reader will recall that Bell Canada had initiated a corporate reorganization in 1982 that resulted in the creation of the holding company Bell Canada Enterprises (BCE) on April 28, 1983. Once the new corporate structure was in place, publishing activities, including print and electronic media, could be run independently of Bell Canada, the regulated company. Robert E. Babe (1990) notes that by 1987 BCE's publishing and printing company, BCE Publi-Tech, had become Canada's premier printing conglomerate. He states further: "By 1987 BCE Publi-Tech, controlling thirty-one subsidiaries, was Canada's biggest printing conglomerate. In June 1988 BCE announced acquisitions of 21 percent of Quebecor Inc., publisher of daily and weekly newspapers in Québec. In exchange, Quebecor was to acquire from BCE Ronalds Printing and certain operations of British American Bank Note Company" (232).

Bell Canada emerged from the enhanced services proceeding relatively unscathed and in a strong position to capitalize on opportunities to provide future enhanced services. Although its competitors had argued before the CRTC that Bell should be required to offer enhanced services through a structurally separate subsidiary, the commission sided with Bell. However, this did not mean that enhanced services would escape regulation. Bell and the other federally regulated carriers would have to offer them on a tariffed basis, subject to CRTC approval. How did the CRTC go about evaluating common carrier requests to offer enhanced services? What criteria did the commission apply in evaluating these requests? What were the perceived obstacles to regulatory approval? It is best to consider these questions by examining a specific case. In what follows we will consider the case of a Bell Canada enhanced service known as ALEX.

8.4 ALEX: THE RISE AND FALL
OF A BELL CANADA ENHANCED SERVICE

On April 13, 1988, Bell Canada applied to the CRTC for approval to begin a market trial of an electronic information and transaction service christened ALEX. The commission responded to the application on May 30, 1988, by inviting public comment on the application, *Bell Canada—Market Trial of ALEX Electronic Information and Transaction Service* (Telecom Public Notice CRTC 1988-22). Interested parties were invited to submit interrogatories to

Bell Canada or comments to the CRTC. During the summer of 1988 Bell responded to the interrogatories. On September 30, 1988, the commission announced its decision to authorize the market trial, *Bell Canada—ALEX Market Trial* (TD CRTC 88-16).

ALEX, modeled to a large extent on the successful French Télétel system, was a gateway service designed for the residential market. Essentially, the ALEX gateway was a directory service designed to simplify interaction between subscribers and information service providers. But it was more than a simple directory; the core of the gateway service was a five-tiered price and billing structure (table 8.1). Each information service was associated with a specific billing tier, chosen by the information provider. The choice of a tier determined both the price of the service and the manner in which the customer would be billed. It also determined how revenues would be distributed between Bell and the information service provider.

The application included a request to offer a dedicated ALEX terminal at rates that were not compensatory. In other words, Bell acknowledged that it would lose money on the terminals. Also the application included a request to offer an electronic version of the telephone white pages.

In considering the application the commission attempted to reconcile the requirements for a successful gateway service with the potential for unfair competition by Bell Canada. The commission was also concerned that the ALEX trial should not entail rate increases to Bell's subscribers: The risks and costs of the market trial would have to be born by Bell Canada's shareholders

Table 8.1   Price Structure for Bell Canada's ALEX Gateway

Category	Charge to Subscriber	Charge to Service Provider
ALEX 1	None	$7.20 /hour ($0.12/min)
ALEX 2	$7.20/hour ($0.12/min) plus possible charges from service provider	No charge, however the service provider may bill directly at its own rate
ALEX 3	$18.00/hour ($0.30/min)	Bell retains $9.00 per hour ($0.15/min) to cover the costs of running the subscriber accounts service ($1.80/hour) ($0.03/min)
ALEX 4	$27.00/hour ($0.45/min.)	Bell retains $9.00 per hour ($0.15/min) to cover network costs of running the subscriber accounts service ($1.80/hour) ($0.03/min)
ALEX 5	Variable. Billed by the service provider	Service provider pays Bell $7.20/hour ($0.12/min); the service provider bills the user directly at a rate it sets for the service

Source: Bell Canada, Guide Préliminaire en vue de devenir fournisseur de service sur ALEX (1988).

and the actual users of the service, and not by subscribers to the monopoly telephone service.

Opposition to the application came primarily from the Centre d'excellence en télécommunications intégrées (CETI) and from the Canadian Business Telecommunications Alliance (CBTA). CETI, a competitor to ALEX, had already launched a gateway service based on French Télétel technology. The service included French Minitel terminals that had been modified for the U.S. and Canadian markets. CETI argued that Bell was taking advantage of its monopoly to enter the market and drive out competitors. Moreover, CETI contended that the scope of the trial was much too large for it to be a considered a market trial. According to CETI, it was not a trial at all, but rather, a full-blown commercial service. CETI also asserted that it was unfair for Bell to be allowed to use its telephone stores (Teleboutiques) to distribute its ALEX terminals, when CETI would not have access to these outlets. Finally, CETI questioned the fairness of Bell being allowed to include promotional information on ALEX in the envelopes it used to bill its telephone subscribers. The CBTA was primarily concerned that Bell would find a way to pass the costs of the trial on to telephone subscribers who would not use or otherwise benefit from the service. The CBTA proposed that Bell be required to offer the service through a separate subsidiary.

In its decision (TD CRTC 88-16), the commission determined that it was not in the public interest to subject Bell Canada to rigorous controls to prevent anti-competitive behavior. The commission appeared to accept the argument that a successful gateway service would only be viable if it were able to attract a large clientele. And this could only be done if, in the initial stages of ALEX development, Bell was able to offer some components of its service on a non-compensatory basis. Similarly, the commission did not wish to limit Bell's ability to market the system. Bell would be allowed to include promotional material for ALEX in the monthly statements it mailed to its subscribers. The commission would not prohibit the distribution of ALEX terminals through its network of Teleboutique stores.

The key regulatory issue was whether Bell should be allowed to operate ALEX in-house, or through a separate subsidiary. In Telecom Decision CRTC 84-18, the commission had made it clear that it was not inclined to impose structural separation on carriers seeking to provide enhanced services. However, in the ALEX trial Bell was seeking authorization for tariffs that were non-compensatory. If ALEX was provided in-house, and on a non-compensatory basis, the commission had to be concerned that Bell would attempt to compensate itself for its losses in its enhanced services through internal transfers from its monopoly telephone business. If left undetected, revenue losses generated in the enhanced service could be carried over to the monopoly telephone business and used to justify rate increases for basic telephone service. This was not primarily an anticompetitive issue, rather it was a consumer issue. It was for this reason that CBTA proposed that ALEX only be offered through a separate subsidiary.

The commission rejected structural separation. In lieu of separation, accounting controls were imposed on Bell Canada. The accounting approach was based on two principles: (1) Bell would have to provide comprehensive information on all the costs and the revenues associated with the trial; (2) the commission would closely monitor costs and revenues in order to ensure that the ALEX trial would not have an undue impact on Bell Canada's revenue requirement (TD CRTC 88-16, 9). That is to say, financial losses due to ALEX could not be carried over to the monopoly telephone business and used to justify rate increases there. Should Bell's rate-of-return fall below the approved level, losses would be passed on to shareholders. Conversely, profits from the ALEX trial could not result in Bell Canada exceeding its maximum approved rate-of-return for any business year.

Bell was unable to gain regulatory approval for its "anchor" service,[3] the enhanced electronic telephone directory. If this service had been approved, it would have included information on a telephone subscriber's profession; thus making it akin to an electronic Yellow Pages. It is not surprising, therefore, that the Canadian Daily Newspaper Publishers Association opposed Bell's request to offer an enhanced white pages. The commission sided with the association, noting that in its previous decision it had ruled "it is the opinion of the [C]ommission that Bell should not be permitted to engage in electronic publishing involving editorial control over content or in the creation or distribution of its own databases" (TD CRTC 84-18, 35). In effect, the commission drew the line at the provision of an electronic white pages (which resembled an electronic Yellow Pages) because of the potential synergy between content and carrier in the provision of such a service. It is interesting to note that Tele-Direct, a subsidiary of Bell Canada, was the publisher of the printed Yellow Pages. The anti-competitive nature of Tele-Direct's exclusive relationship with its parent company, Bell Canada, was not problematic for the commission. This was the case because Tele-Direct was working in another media, print, and was operating as a separate subsidiary. This was sufficient, in the commission's view, to place this relationship outside the restrictions of the Bell Canada Act.

On December 30, 1993, Bell Canada applied to the CRTC for permission to withdraw the ALEX service. The company cited lack of interest on the part of information providers and consumers as the motive. The service had never turned a profit and it was unlikely that it ever would, *Bell Canada—Application to Withdraw ALEX Service* (TD CRTC 94-4).

8.5 FROM ENHANCED SERVICES TO LOCAL NETWORK COMPETITION: A POLICY FOR TELEPHONE/CABLE CONVERGENCE

We have seen that the regulatory treatment of enhanced services has been an important factor in determining when, and under what conditions, new services will be brought to market. But the case of enhanced services is significant in another respect. At a more fundamental level, it reveals the difficulties inherent in creating and maintaining boundaries between services and, indeed, between entire industries, where there is asymmetrical regulation—that is to say, when the companies that provide the services are subject to different degrees of regulation. In the Canadian model, common carriers offering enhanced services were subjected to regulation due to the requirement that services had to be offered on a tariffed basis. Non-common carriers did not have to have their rates approved by the commission.

In the case of enhanced services, the decision to allow common carriers to provide these services presented regulators with a unique set of problems. On the one hand, prohibiting the regulated common carriers from operating enhanced services could delay the introduction of new services, or result in them not being introduced at all. On the other hand, allowing the regulated companies to offer services in competitive markets was a threat to competitors and potentially unfair to ratepayers of the monopoly service.

The final solution to this regulatory dilemma would be to dismantle what remained of the common carriers' monopoly by liberalizing the local network market. This approach was first proposed by the FCC in 1986, when it introduced the concept of Open Network Architecture as part of its *Computer III* rules. If a workable framework could be developed to introduce competition to the last network monopoly, local networks, then there would be no reason to continue to maintain the traditional barriers to entry—including barriers that had discouraged cable TV companies from offering telephone service. In fact, there was a perception that competition in local networks would be sustainable if both the telephone and cable companies were allowed to enter each other's traditional markets.

Cable TV networks and telephone networks have traditionally been subject to separate regulatory treatment. The cable companies were regulated under the Broadcasting Act. The telephone companies were subject to regulation under the Railway Act. The cable TV companies were regulated as broadcast distribution undertakings. The telephone companies were regulated as common carriers. This separate regulatory treatment overlooked the fact that both industries operate network facilities that provide wire-based connections directly to the residential market.

The communication infrastructures of Canada and the United States are unique due to the fact that in both countries one finds two mature industries providing *wired-based* connections to the residential market. In other countries, cable TV has been introduced much later, and has had to compete directly with another distribution technology, direct-to-home broadcast satellites. As a result, penetration rates for cable networks elsewhere in the world are much lower. The existence of a dual-wire infrastructure has presented a unique opportunity to the CRTC and the FCC as they attempt to extend competition to local networks. The challenge for policy makers has been to preserve and build upon the dual-industry structure. On the one hand this means creating policy that will encourage industry crossovers: cable entry into the telephone business; telephone entry into the business of program distribution. On the other hand, policy makers must prevent cross-industry takeovers: telephone company purchases of cable facilities within the same service territory or vice versa. If this were to occur, it would put an end to any hope of real facilities competition at the local level.

8.6 THE NEW REGULATORY FRAMEWORK, PART I: RATE REBALANCING, CROSS-SUBSIDIES, AND THE END OF RATE-OF-RETURN

In the early 1990s the CRTC began proceedings that would prepare the way for the introduction of competition to local networks. Following closely after the decision to liberalize long distance (TD CRTC 92-12), the CRTC announced on December 16, 1992, that it would begin a public hearing to consider whether changes should be made to the regulatory framework in order to expand competition in the telephone market, *The Review of the Regulatory Framework* (Telecom Public Notice CRTC 92-78). The CRTC noted the following: "However, while the telephone companies are operating in an increasingly competitive environment, and thus may be subject to a greater degree of market discipline, *they continue to maintain effective control of the provision of network access and local services* and to dominate the public long distance market" (paragraph 2, emphasis added).

The path toward full liberalization, including local networks, was not a straight one. The commission would first have to settle a number of issues that had not been resolved in the decision to liberalize long distance. These included: rate rebalancing, cross-subsidies (contributions to local telephone service), and the future of rate-of-return regulation.

It is typical for monopolists to call for rate rebalancing following a decision to liberalize long distance. The term "rate rebalancing" is a regulatory euphemism for increasing rates for local service in the wake of a decision to liberalize long distance. The introduction of competition to long distance markets has the effect of driving down the price of long distance; this, in turn, will reduce the subsidy (contribution) that flows from long distance to local service. Monopolists argue that local service rates must rise in order to make up for the lost subsidy. Although rate rebalancing was not part of the decision that introduced competition to long distance, the CRTC did not reject the possibility that some form of rate rebalancing might be necessary at some point in the future. However, the CRTC had never fully endorsed Bell Canada's estimates of the amount of the subsidy. Recall that the CRTC had concluded in its modified version of the

Five-Way Split in 1985 (*see* chapter seven) that a number of service categories were not making a fair contribution to the operation of local networks.

In Telecom Decision CRTC 94-19, *Review of the Regulatory Framework,* the CRTC undertook a comprehensive restructuring of the way rates would be set for telephone services in the future. The restructuring began with the following initiatives:

1. Rate increases that would see monthly rates for local service increase by $2 during each year of a three-year period. At the end of the three-year period, rates for local service would have risen $6 a month for a total yearly increase of $72[4]
2. The contributions (subsidy) to local service paid by all long distance carriers (including Bell Canada and the other Stentor members) would be reduced
3. A new tariff, know as the Carrier Access Tariff (CAT) would be introduced for all carriers; its purpose would be to recover all the underlying costs and charges associated with access to the local network. The CAT would compensate for the reduction in the contribution charge paid by long distance companies; unlike the contribution charge it would apply to other services that used the local network (II, B. 2)

These changes to the regulatory framework were designed to bring rates for local service closer to actual costs, and to ensure that other services that made use of the local network would contribute fairly to its operation. By raising rates, by reducing the subsidy that flowed from long distance to local access service, and by ensuring that all long distance carriers would make contribution payments, the commission would achieve a number of objectives. First, it would ensure that the former monopolistic carriers would not be unfairly disadvantaged when competing in the newly liberalized long distance market because of a contribution obligation which exceeded that of its competitors (AT&T Canada, Sprint, and so on). Second, maintaining contribution payments in support of local service would ensure that local access service would continue to be affordable. Finally, the combination of rate increases and continued contribution payments would bring revenues for local service closer to actual costs, thus clearing the way for a rational introduction of competition to the local access market. If costs in the local exchange market exceeded revenues by a wide margin, then market pricing set under the new competitive regime would not be based on accurate information and would be uneconomic.

The CRTC also announced in Telecom Decision CRTC 94-19 that it intended to eliminate earnings-based regulation on January 1, 1998. The rate-of-return formula, which had been the cornerstone of monopoly regulation for the better part of a century, would be superseded by price caps on local service. This would achieve two objectives: (1) it would eliminate incentives for the Stentor companies (formerly Telecom Canada) to conceal cross-subsidies within their general revenue requirements; and (2) it would allow the Stentor companies more flexibility in setting rates for individual services. Because the change to price caps would not be immediate, the commission introduced an interim framework designed to split the telephone companies' business into two segments: utility (monopoly) and competitive. The utility (local access and local exchange) services would continue to be regulated under the rate-of-return formula. However, as noted earlier, this included a pre-authorized rate increase for local service. Revenues from the competitive segment (including long distance) would no longer be included as part of the calculation to determine the revenue requirements of the telephone companies. This was an interim framework, pending a decision on the rules that would govern the full liberalization of the local telephone market.

On March 12, 1996, the CRTC announced that it was initiating a proceeding entitled, *Price Cap Regulation and Related Issues*, to determine the form of price cap regulation (Telecom Public Notice CRTC 96-8). The price cap regime for local service would apply to the following companies (all members of the Stentor consortium):

B.C.Tel
Bell Canada
The Island Telephone Company (Island Tel)
Maritime Tel & Tel (MT&T)
MTS NetCom (MTS; formerly Manitoba Telephone System)
The New Brunswick Telephone Company (NB Tel)
NewTel Communications (formerly Newfoundland Telephone Company)
TELUS Communications Incorporated (TCI)

Evidence and proposals for the price cap regime were received from the telephone companies and from other interested parties. The CRTC issued its decision on May 1, 1997, *Price Cap Regulation and Related Issues* (TD CRTC 97-9).

In its decision, the commission summarized the rationale behind the shift from rate-of-return to price caps in the following manner:

> Several factors inherent in price caps make it a more efficient and effective form of regulation than rate base/rate-of-return regulation. The traditional method, focusing on earnings on invested capital tended to encourage investments, creating a need for ongoing regulatory scrutiny of capital expenditures, expenses and earnings. The price cap mechanism, on the other hand, focuses on prices, which are of direct concern to subscribers, and creates incentives for the telephone companies to become more efficient, subject only to the constraints of the regime. The regime requires telephone companies to first meet a productivity target set by the [C]ommission, which benefits subscribers by keeping prices down, and then, by exceeding this productivity target, companies can reap benefits for their shareholders. (Canadian Radio-television and Telecommunications Commission 1997)

The price cap regime would take effect January 1, 1998, and would be in effect for four years. Before the price cap plan expired in 2002, the CRTC would review the plan to determine if the parameters had been set correctly. Price caps would apply to basic residential and business local service provided by the telephone companies. In the case of basic residential service (local exchange service), rate increases would be limited to the rate of inflation (as determined by a price cap index [PCI]) minus an adjustment for anticipated productivity gains by the telephone companies, known as the (productivity offset). The productivity offset would be set at 4.5 percent. The inclusion of a productivity offset would ensure that rate increases would be lower than the rate of inflation.

In order for competition in local networks to be viable, competitors must have access to the incumbent carriers' local networks. It was the CRTC's task to determine how interconnection would occur. Networks would have to be *interoperable*, that is, technically compatible with one another. This meant that competitors and incumbents would have to agree on technical standards. At another level was the problem of actual physical interconnection of facilities. Networks would have to be interconnected so that customers of one service provider would be able to complete calls to customers subscribing to another service. Rules concerning where and how the *interconnection* of networks would take place would have to be established by the commission. The commission would have to decide whether competitors should be allowed to install equipment on the incumbent's premises, that is, central offices. This was known as the problem of *co-location* of facilities. In order to avoid uneconomic duplication of facilities, incumbents would be required to unbundle certain components of their local networks and make them available to competitors. The commission would have to determine which components would be *unbundled* and how they would be priced. It was also clear that for local competition to succeed, subscribers would have to be allowed to keep their old telephone numbers when switching to a different local exchange provider. This was known as the problem of *number portability*.

8.7 THE NEW REGULATORY FRAMEWORK, PART II: INTEROPERABILITY, INTERCONNECTION, CO-LOCATION, UNBUNDLING, AND NUMBER PORTABILITY

In Telecom Decision CRTC 94-19, *Review of the Regulatory Framework*, the commission identified *interoperability, interconnection, co-location, unbundling*, and *number portability* as issues that would have to be dealt with during the transition period leading up to the full liberalization of local networks. In other words, the regulatory framework decision established a general policy objective: the introduction of competition to local telephone networks. How this would be accomplished would be the subject of consultation and negotiation with the incumbent telephone companies, potential competitors, and users (residential and business). The general direction and destination was clear to all the actors. Only the modalities of how competition would occur remained to be worked out.

Telecom Decision 94-19 resulted in the price cap plan described earlier; but it also produced a landmark decision on local competition. On the same day the details of the price cap plan were announced (May 1, 1997), the commission issued its decision on *Local Competition* (TD CRTC 97-8). The process that led to the local competition decision began on March 15, 1995, when, as prescribed in Telecom Decision CRTC 94-19, the Stentor companies submitted a proposal for a model tariff for the interconnection and unbundling of local networks. The CRTC responded on July 11, 1995, when it initiated an inquiry entitled, *Implementation of Regulatory Framework—Local Interconnection and Network Component Unbundling* (Telecom Public Notice CRTC 95-36). An oral hearing was conducted from August 19 to September 4, 1996, to receive comments and counter proposals to the Stentor plan. Testimony was heard from over twenty-five parties, including the following: the Stentor Companies, AT&T Canada Long Distance Services Company, Sprint Canada, the Canadian Cable Television Association, the Consumer's Association of Canada, the Director of Investigation and Research (Bureau of Competition Policy), and several provincial governments.

The decision to open the local exchange to competition resulted in a highly detailed ruling that covered topics such as how traffic for 1-800 and 1-888 would be directed; how local exchange operators would be compensated for the costs of terminating traffic; whether competitive local carriers should be required to provide 911 service under tariff, and so on. Limitations of space prohibit a comprehensive analysis of the decision. Essentially, the decision dealt with two issues: interconnection and unbundling. A key to the CRTC's pro-competitive framework was the relationship between the existing telephone companies (referred to in the decision as *incumbent* local exchange carriers [ILECs]), and potential competitors (referred to as *competitive* local exchange carriers [CLECs]). The framework prescribed obligations and controls based on whether a local exchange carrier was an ILEC or a CLEC.

The goal of interconnection was to create a network of networks. If the policy was successful, there would be multiple local exchange carriers operating in the same service territory. So for example, two subscribers living on the same street would be able to receive service from two different local exchange carriers. Typically, one of them would be an ILEC, the other a CLEC. The interconnection of these networks would have to be seamless so that subscribers to one local exchange carrier would be able to complete calls to subscribers of another local exchange carrier, regardless of whether they lived across town from one another, or across the street. At a technical level, this meant that there would have to be standard interfaces. The commission would have to determine how costs would be shared for the trunks that linked exchanges together. It would have to set rules for the interconnection of the signaling networks, which performed various call processing functions (including communicating with databases to determine how calls are routed). These and many other issues were dealt with in Telecom Decision CRTC 97-8 (primarily in section II of the decision).

Interconnection policy deals with the relationship between local service providers at the level of the local switch (exchange). It presupposes that interconnection will take place between two or more local service providers all of whom own switching equipment. At this level of operation, there is a certain equality in the relationship between ILEC and CLEC: minimally, they are both equipped with switching platforms. However, this symmetrical relationship breaks down as one moves from the switch to the local access component of local service. It is unlikely that a CLEC would be able to duplicate all the facilities and services that currently comprise an ILEC's local exchange service and still be competitive.[5] The CRTC's competitive framework takes this into account by requiring the ILECs to *unbundle* a number of components that are basic to the provision of local service. These unbundle components can then be acquired by a CLEC and mixed with its own facilities as part of a local exchange service.

Telecom Decision CRTC 97-8 requires the mandatory unbundling by the ILECs of *essential facilities*. What is an essential facility? This question is answered in section IV of the decision, where the commission defines essential facilities and identifies which facilities will be considered essential. The decision stipulates that a facility, function, or service will be considered essential if it meets the following three criteria: "(1) it is monopoly controlled; (2) a CLEC requires it as an input to provide services; and (3) a CLEC cannot duplicate it economically or technically." Using these criteria, the parties to the hearing generally agreed that the following services or functions would be considered essential: central office codes, subscriber listings, and local loops (in certain service areas). Central office codes are the first three digits of a local telephone number. Subscriber listings are the database containing the subscriber's name, address, and telephone number. Local loops are the transmission paths connecting a subscriber's premises to the main distribution frame in an ILEC central office.

There was a consensus that local loops in rural areas, and certain urban areas with low population density, should be considered essential facilities. Due to the low density of subscribers in these areas, costs per line may be considerably higher than in urban areas. This means that there are few incentives for CLECs to duplicate local loop facilities in these areas. In order to grasp the economics of wire-line facilities in rural areas, one only has to consider the current state of cable TV penetration in these areas. Cable TV is unavailable in many rural communities due to the high cost of providing local loops. As a result, rural areas have proven to be a fertile market for satellite TV services.

The commission was not convinced that local loops in areas with greater subscriber density (such as urban centers) constituted essential facilities. However, because it considered it unlikely that the CLECs would be able to provide a significant number of loops in these areas during the early years of competition, it was prepared to designate these facilities as essential for a period of five years. Once the five-year period had expired, there would be no mandatory requirement that ILECs unbundle local loops in urban areas. On the other hand, the commission determined that local switching, directory assistance databases, directory assistance services, ILEC white pages directories, access to rights-of-way, and 911 service *did not* constitute essential services. The pricing of essential services would be based on costs plus a 25 percent markup. Costs would be calculated using an already agreed to accounting procedure known as Phase II. The 25 percent markup was designed to compensate for fixed common costs that were not included in the Phase II formula.

On June 16, 1997, the CRTC issued its decision on the co-location of facilities, *Co-location* (TD CRTC 97-15). An approach to number portability would be developed on a cooperative basis by an industry working group under the auspices of the CRTC Interconnection Steering Committee (TD CRTC 97-8, paragraph 10).

The decisions on local competition (TD CRTC 97-8) and price caps (TD CRTC 97-9), both issued on May 1, 1997, cleared the way for local telephone competition on January 1, 1998. The decisions complete a policy process that was initiated with Telecom Public Notice CRTC 92-78, *Review of the Regulatory Framework,* and took shape in TD CRTC 94-19, *Review of the Regulatory Framework* (tables 8.2 and 8.3). Approximately six years after the decision to liberalize the long distance market, a framework was in place that would eliminate the last remnant of the telephone companies' vertically integrated monopoly. From a regulatory standpoint, the telephone companies' monopoly over the provision of local telephone service ceased to exist on January 1, 1998. However, the new regulatory framework is not a guarantee that competition will occur. One cannot dictate competition by regulatory decree. Nor does the pro-competitive framework ensure that the CLECs will be able to break the telephone companies' domination of the local exchange market. The success of the CRTC's policy will not be gauged in terms of whether the ILECs have adapted in good faith to the new rules. It will be measured by the availability of alternative local service in both residential and business markets, and the overall affordability of basic telephone service in all service areas.

8.8 A POLICY FOR TELEPHONE/CABLE CONVERGENCE

The new, more competitive regulatory framework would be a two-way street. The incumbent telephone companies would see their local network monopo-

Table 8.2   Evolution of CRTC Competition Policy

Telecom Decision CRTC 92-12 12 June 1992	*Competition in the Provision of Public Long Distance Voice Telephone Services and Related Resale and Sharing Issues*
Telecom Decision CRTC 94-19 16 September 1994	*Review of the Regulatory Framework*
Telecom Public Notice CRTC 96-8 12 March 1996	*Price Cap Regulation and Related Issues*
Telecom Decision CRTC 97-9 1 May 1997	*Price Cap Regulation and Related Issues*
Telecom Decision CRTC 97-8 1 May 1997	*Local Competition*

Table 8.3   Evolution of CRTC Competition Policy in the Wake of Telecom Decision CRTC 94-19

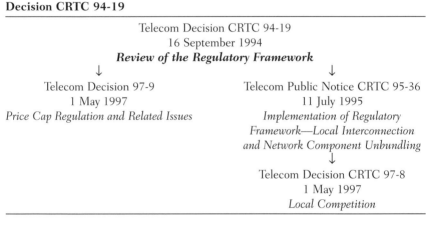

lies opened to competition, but what would they receive in return? As was the case with the FCC's approach to enhanced services, Canada's new regulatory framework is based on a quid pro quo. In exchange for relinquishing the last vestiges of their monopoly (the local network), the incumbent telephone companies will be allowed to participate in markets for programming distribution, programming production, and other content-based services. However, in order for this to occur, significant changes had to be introduced to broadcast regulation and broadcast policy as it pertained to the telephone companies. During the period of monopoly regulation, the telephone companies were prohibited from engaging in broadcasting activities. In order to fully appreciate the extent of change, it is useful to consider how the telephone and cable companies were regulated under the old regime.

Canada's Broadcasting Act identifies three types of broadcasting undertakings: distribution undertakings (i.e., cable TV), programming undertakings, (i.e., The Sports Network), and networks (CTV and Canadian Broadcasting Corporation) (2 (1)). In order to operate a broadcasting undertaking in Canada, a license is required from the CRTC. Therefore, a telephone company seeking to act as a broadcasting undertaking would require a license from the CRTC. In the case of Bell Canada, the prohibition on holding, or even applying for, a broadcasting license was contained in the Bell Canada Act, as revised in 1967–1968. The Manitoba Telephone System and Saskatchewan Telephones were specifically prohibited from holding a broadcasting license as a result of an Order in Council issued in 1972, which stipulated that provincial crown corporations were not eligible to hold broadcast licenses (Department of Communications 1992, 95). In 1969 the CRTC issued a public announcement entitled "Licence Policy in Relation to Common Carriers," which stated that it would not be in the public interest for common carriers to hold cable TV licenses (95). The official justification for the prohibition was the long-standing common carrier principle that content and carriage should be separated in order promote equity and discourage discriminatory or preferential practices. This policy was reinforced by the CRTC's decision to promote growth and stability in the cable industry so that the industry could serve as a distribution outlet for indigenous Canadian programming.

Because the cable companies operated as broadcasting distribution undertakings, and not common carriers, their ability to diversify their operations was governed by a different, more liberal set of rules. In 1971 the CRTC, in its first statement of cable policy, encouraged cable companies to participate in the development of non-commercial, local programming on community channels. Over the years, the CRTC has allowed cable companies to participate in the production of programming content; but only on a case-by-case basis and not as a matter of general policy. Also on a limited case-by-case basis, the commission has sanctioned cable company ownership of over-the-air radio and TV stations (Department of Communications 1992, 117). Non-programming services, such as security/alarm services have not been regulated by the CRTC, because they do not qualify as broadcast activities under the Broadcasting Act and, therefore, are not subject to regulatory scrutiny under the act. This has meant that there were no rules to prohibit cable companies from discriminating against other providers of non-programming services should they wish to use the cable network to provide a service. On a voluntary basis, the cable industry did establish accounting guidelines to ensure that non-programming services were not financed at the expense of ratepayers to their monopoly programming distribution operation. But these were not as rigorous as those that applied to the telephone companies (118).

Essentially, participation in markets for TV programming will mean three things for the telephone companies: (1) the provision of broadband networks on a common carrier basis, (2) the right to distribute programming (like the cable companies) on broadband platforms, and (3) the right to produce programming

content. Under the common carrier model, the telephone company would simply provide bandwidth to other companies that would be responsible for programming decisions: what channels to carry, pricing, service tiers, and so on.
Under the programming model, the telephone companies would control both
network facilities and programming decisions, much like a cable company.
When the CRTC announced the *Review of the Regulatory Framework* in 1992
(Telecom Public Notice CRTC 92-78), it did so under its authority to regulate
telecommunications. This meant that the commission could not consider telephone company entry into the business of broadcasting programming distribution, since this was under the Broadcasting Act. It would have to be dealt with
in a subsequent proceeding.

Within the framework of the Telecommunications Act (1993), the commission was, nevertheless, able to address a related video distribution technology known as video dialtone (VDT). VDT is the broadband[6] platform required to deliver a video-on-demand (VOD) service. VOD resembles an
electronic version of a video store that rents cassettes. A subscriber to VOD
would be able to select films or other programs from a video catalog (database); these would then be delivered using some form of compression technology to a decoder box attached to the television set. Unlike pay TV there is
no programming schedule: a film can be selected and delivered for viewing at
any time of the day. In Telecom Decision CRTC 94-19, the CRTC determined that telephone companies should be allowed on *a common carrier basis*
to provide a VDT service. In other words, they could provide platforms (bandwidth) for VOD and other broadband services (II. E.4). The commission defined VDT narrowly as the "technological capability through which broadband services may be offered" (II. E.4). The definition did not encompass the
programming (content) that would be delivered through the broadband network. In other words, the decision did not include a ruling on whether the
telephone companies should be allowed to operate a VOD service. But it did
stipulate that VOD constituted a broadcasting undertaking under the Broadcasting Act and would have to be offered by providers that had been licensed
by the CRTC.

The commission would not have to wait long for an opportunity to formulate policy under the Broadcasting Act. In the spring of 1994 Canada's industry minister announced that the government would soon engage in a nation-
wide consultation as a first step towards creating a national policy for the
development of Canada's Information Highway (table 8.4).[7] To this purpose,
an Information Highway Advisory Council (IHAC), chaired by former McGill
University principal David Johnston, was formed. The council was comprised
of twenty-nine members and was charged with submitting a report to the government that would include recommendations on how to develop Canada's Information Highway.[8] On October 11, 1994, the government issued Order in
Council P.C. 1994-1689, entitled *Call for Comments Concerning Order in*

Council P.C. 1994-1689 (Telecom Public Notice CRTC 1994-130), which requested the CRTC to report on "a number of wide-ranging questions covering three broad areas: facilities, content and competition." The CRTC's report would be submitted to the IHAC for comment. The government would consider both reports before formulating policy on the Information Highway.

In response to the Order in Council, the CRTC issued a public notice calling for comments. Following a public consultation process the CRTC issued its report on May 15, 1995. In the report entitled, *Competition and Culture on Canada's Information Highway: Managing the Realities of Transition* (Canadian Radio-television and Telecommunications Commission 1995), the CRTC outlined for the first time how the new regulatory framework would deal with telephone company applications to hold licenses under the Broadcasting Act.

The commission began by stating that it was prepared to grant broadcasting distribution licenses to the telephone companies so that they could offer VOD on VDT platforms (Canadian Radio-television and Telecommunications Commission 1995, 2.1.b). But this would only occur once the telephone companies had filed non-preferential tariffs for the basic VDT service (the common carrier broadband platform). In other words, the telephone companies would be authorized to operate VOD, a distribution service (essentially, the catalog [database] of films, documentaries, and other programs) via its own VDT platforms. The telephone companies could apply for licenses to provide broadcast programming (i.e., to create programming content); however, these licenses would have to be held by a structurally separate entity.

The CRTC's approach to future broadband cable networks was based on the same principles that had underscored its approach to broadband tele-

Table 8.4   Evolution of Canadian Information Highway Policy

Intervenor	Event	Date
Industry Canada	Beginning of Public Consultation on Canadian Information Highway Policy; creation of the Information Highway Advisory Council	Spring 1994
Federal Cabinet	Federal government mandates the CRTC to submit a report on Canada's Information Highway	October 11, 1994
CRTC	Telecom Public Notice CRTC 1994–130, *Call for Comments Concerning Order in Council P.C. 1994–1689*	1994
CRTC	*Competition and Culture on Canada's Information Highway: Managing the Realities of Transition*	May 15, 1995
Industry Canada	*Convergence Policy Statement*	July 10, 1996

phone networks in Telecom Decision CRTC 94-19. The strict application of
the principle of separation of content and carriage would be renounced.
Cable operators would be allowed to provide specialty programming channels
over their cable networks on the condition that the network would also be accessible to other service providers. However, before a cable company could
offer a specialty programming channel, two conditions would have to be met:
(1) there would have to be enough channel capacity to allow other service
providers to use the cable broadband platforms; (2) there would have to be
rules to ensure that the other providers would have equal access to the cable
network facilities. Before a cable company could be licensed to provide a
VOD service a further condition would have to be met: the company would
have to establish non-preferential common carrier VDT tariffs (Canadian
Radio-television and Telecommunications Commission 1995, 2.1.(d)). As
was the case with the telephone companies, all broadcasting programming
services would have to be held by a structurally separate entity. If these conditions were met, then the cable companies would be allowed act as both network operators and programming providers.

As was the case with the telephone industry, the approach to cable networks
was based on a quid pro quo. The cable companies would be granted licenses
to provide content for their networks on the condition that they transform
themselves into common carriers. In the case of cable companies, it is far from
certain that the quid pro quo will work. It remains to be determined whether
opportunities for profit in programming production will be sufficient to outweigh the burdens associated with becoming a common carrier.

On July 10, 1996, the Canadian government issued its *Convergence Policy
Statement* (Industry Canada 1996). The statement culminated the public
consultation process that had begun with Order in Council P.C. 1994-1689.
Before issuing the statement, the government considered the reports submit-
ted by the CRTC (which has already been discussed) and the IHAC, entitled
Connection, Community, Content. The Challenge of the Information Highway
(Industry Canada 1995a). Essentially, the government endorsed the approach
to telephone/cable convergence articulated by the CRTC in 1994 and 1995.[9]
The government announced that it would take the following steps that would
clear the way for the telephone companies to hold broadcasting licenses:

> All telecommunications carriers that meet the Canadian ownership and control
> requirements will be eligible to hold broadcasting licences. Amendments to the
> Bell Canada Act will be introduced to eliminate the prohibition from holding
> broadcasting licences.
>
>   The Order in Council which limits the ability of crown corporations to hold
> broadcasting undertakings will be modified to allow such corporations to hold
> broadcasting licences while recognizing the obligation for broadcasting licensees
> to operate at arm's length from governments, including programming independ-
> ence. (Industry Canada 1996, 5)

Moreover, the government was prepared to remove the prohibition on B.C.Tel and Québec-Téléphone holding broadcasting licenses. Under the Broadcasting Act, all broadcasting undertakings operating in Canada had to be Canadian owned and controlled. B.C.Tel and Québec Téléphone are subsidiaries of the American telecommunications company GTE. With respect to these two incumbent telephone companies the government was prepared to grant an exemption (Industry Canada 1996, 5).

To summarize, under the new framework the telephone companies would be allowed to act as broadcasting distribution undertakings, under license, and without resorting to separate subsidiaries. They would also be granted licenses to act as originators of programming content, under license, on the condition that these operations are run by structurally separate entities. However, the right to hold broadcasting licenses will be contingent upon the CRTC determining that their local networks have been open to competition. Finally, they will be able to provide non-programming services without recourse to a license or separate affiliates. A similar framework will apply to cable companies. If the cable companies are able to satisfy the CRTC's criteria of non-discriminatory access to their broadband networks, they will be granted licenses to produce programming content for these networks. However, this will have to be done through separate subsidiaries. Finally, the cable companies will continue to be able to provide non-programming services via their networks.

The new regulatory framework is designed to encourage the telephone companies to invest in broadband networks by allowing them to benefit from investment in a new role as broadcasting distribution undertakings and broadcasting programming undertakings (albeit subject to separate subsidiary requirements). But the framework is not just about developing facilities, it also embodies objectives related to the promotion of Canadian programming. These objectives have been enshrined in the Broadcasting Act since 1968. The most recent affirmation of this purpose was expressed in the 1991 revised Broadcasting Act, which states that the broadcasting system should:

> encourage the development of Canadian expression by providing a wide range of programming that reflects Canadian attitudes, opinions, ideas, values and artistic creativity, by displaying Canadian talent in entertainment programming and by offering information and analysis concerning Canada and other countries from a Canadian point of view, . . . each element of the Canadian broadcasting system shall contribute in an appropriate manner to the creation and presentation of Canadian programming; . . . each broadcasting undertaking shall make maximum use, and in no case less than predominant use, of Canadian creative and other resources in the creation and presentation of programming, . . . the programming originated by broadcasting undertakings should be of high standard. (3.(1)(d) (ii-iv))

The new framework will promote the objectives of the Broadcasting Act by requiring the new broadcasting undertakings to respect the obligations that currently apply to Canada's programming undertakings and cable companies. For example, Canada's over-the-air broadcasters and off-the-air specialty programmers must respect content and spending requirements designed to give support to indigenous programming. Cable operators (programming distribution undertakings) must give priority to local and regional signals. There are limits on the number of non-Canadian channels that can be carried. When approving applications for specialty channels, the CRTC has attempted to ensure that there is a balance between Canadian and U.S. offerings on cable tiers. Cable operators must contribute to a fund for Canadian programming.

The pro-competitive framework distinguishes between two categories of programmers: programmers who offer a "scheduled sequence of programming" and programmers whose programmes are available "on demand," such as VOD. The first category of programmers will be subject to Canadian content and spending requirements that already apply to specialty programming, pay, and pay-per-view services. VOD services will also require a broadcasting license. As a condition of license, VOD providers will have to include a "maximum practicable number of Canadian titles" and they will have to organize the menu of titles to give prominence to Canadian productions. The new programming distribution undertakings and VOD providers will be required to make a financial contribution to the Canadian programming fund. The amount of the contribution, which will also apply to the current cable companies, will be a percentage of gross revenues. Similar obligations will also apply to direct-to-home satellite undertakings.

8.9 CONCLUSION

Following the decision to liberalize markets for long distance telephony, the CRTC moved slowly but deliberately towards full liberalization of the telephone industry. The process took over six years to complete. It began with the announcement on December 16, 1992, that the commission would begin hearings to review the regulatory framework (Public Notice CRTC 92-78). The process culminated in two decisions announced May 1, 1997: Telecom Decision CRTC 97-8 (which established the rules for the introduction of competition to local networks); and Telecom Decision CRTC 97-9 (which outlined the price cap framework that would apply to local rates). Between these two events the new policy orientation was articulated in two documents: a regulatory decision (TD CRTC 94-19) and a CRTC report prepared for the government in conjunction with Information Highway project (CRTC 1995). The approach the commission had taken towards local competition and cable/telephone convergence was confirmed by the government on July 10, 1996, when it issued its *Convergence Policy Statement* (Industry Canada 1996).

The events we have described in this chapter do not stand in isolation. The opening of local telephone markets to competition on January 1, 1998, culminated a process that began with the Telecommission studies of the late 1960s (Department of Communications 1971a) and the publication of the government's first Green Paper on telecommunications in 1972 (Department of Communications 1973). The CRTC's decision to encourage competition in local telephone markets marks the end of a century of vertically integrated *monopoly* in the telephone industry. More precisely, it marks the end of government and regulatory agency *support for* a vertically integrated monopoly. It is now official policy to promote competition wherever possible in the telecommunications industry. But will the end of government support for vertical monopoly lead to true competition in all sectors of the telephone industry? It is one thing to create a framework for competition, it is quite another to create a *successful* framework. It is one thing to remove barriers to entry in the industry, it is quite another to identify and create the conditions that will provide adequate incentives (opportunities for profit) for entry.

What lessons may be learned from this final phase in the liberalization of the telephone industry? First, it is clear that under the new regime the alternative to monopoly is not freewheeling competition, but a regime based on *pro-competitive regulation*. Regulation designed *to promote competition* has replaced regulation designed *to protect the monopolist from competition*. Second, we have seen that the new pro-competitive framework is based on a complex set of rules that are quite extensive and often highly detailed. Market liberalization has not led to deregulation, understood as the withdrawal of regulatory oversight and control, but rather to a new mode of control (a new set of rules), albeit geared to a different set of pro-competitive objectives. Third, we have seen that the decision to open the local network market to competition is not an end in itself. It is best understood as an *instrumental policy* designed to advance another set of policy objectives related to the development of Canada's so-called Information Highway. This means that competition in the local network market is designed to foster investment in Canada's telecommunications infrastructure. If the policy is successful, Canada's facilities-based providers will invest in broadband networks that will deliver new services to the residential market. If the telephone and cable industries do not invest in local broadband networks, it will be because the policy has failed. Finally, the new regulatory framework does not constitute a complete break with the past. The framework maintains two policy practices associated with the old regime: the *continuance of subsidies to local networks* in order to maintain universality (albeit at a lower level), and the *support for the Canadian programming industry* in the form of regulation and mandatory contributions to a Canadian programming fund.

To summarize, the new regulatory framework presents a contradictory picture. It promotes competition, not by removing regulation, but by giving regulation a new pro-competitive orientation. Although it represents a radical departure from the past, it incorporates two of the most controversial aspects of the old regime: network subsidies and a complex set of rules and subsidies in support of the Canadian programming industry. Although the framework complies with current economic orthodoxy concerning the power of markets to promote technological innovation, it is still largely untested as an approach to promoting investment in network infrastructure at the local level. Although the Canadian approach to liberalization has been more orderly than its American counterpart, it is still a vast experiment, with few precedents in the history of the telecommunications industry. Ultimately, the success of the project will not be judged on the basis of the process that produced the change, but rather on its results. The real test of the policy will not just be whether there are competitive sources of local exchange service, but whether local networks have the capacity to deliver true broadband services that are accessible and affordable to the residential market.

NOTES

1. Common carriers use market trials to evaluate consumer interest in a new service. Market trials are done on a small scale in a limited service area. They enable a company to evaluate what economists refer to as the "demand elasticities" for a product or service: the relationship between price and demand. They also enable companies to get a better idea of the true costs of providing a service. Presumably, at the end of a market trial a company will have a clearer idea of whether the service will be profitable. A market trial may be preceded by a "technical" trial, and if successful will be followed by a full-scale commercial offering of the new service. Regulatory approval is necessary to assure that service does not require subsidies from the common carrier's monopoly operations for it to be financially viable.

2. The Canadian Industrial Communications Assembly, the Canadian Association of Data and Professional Service Organizations, the Canadian Bankers' Association, and Datacrown Incorporated (TD CRTC 85-17, 2).

3. An anchor service is service with high utility for subscribers. It is designed to bring subscribers to the system where they will discover, and potentially, subscribe to other services.

4. In response to consumer pressure, the CRTC temporarily reversed the automatic $2-a-month increase in the third year. However, in Telecom Decision CRTC 97-9, *Price Cap Regulation and Related Issues,* which dealt with the new price cap regime, the CRTC authorized a final $3 increase effective January 1, 1998 (paragraph 23).

5. The local loop that links a subscriber's premises to the central office (the first switch) is an example of a component that may not be duplicated on an economic basis.

6. Cable TV is an example of a broadband service. Voice telephony is an example of a narrowband service. The difference stems from fundamental differences in the network facilities they use. A broadband network has sufficient capacity to carry voice, high-speed data, and video signals. By definition, networks based on coaxial cable and fiber optics are broadband. Networks that employ paired copper wires are narrowband. However, it is possible to use data compression technology to boost the channel capacity of a copper wire network so that it has some of the functional capabilities of a broadband network.

7. The term "information highway" is essentially a rhetorical device. In practical terms it refers to future *broadband networks* and the services they will carry. Unlike today's cable networks, these will be switched networks, thus allowing subscribers to both send and receive information. Unlike today's switched telephone networks, they will have the channel capacity to transmit video images.

8. The Information Highway Advisory Council (IHAC) submitted its report to the government in September 1995 (Industry Canada 1995a).

9. Telecom Decision CRTC 94-19, and the Canadian Radio-television and Telecommunications Commission 1995.

Conclusion to Part II

In the preceding chapters we have chronicled the transition from monopoly to competition in the Canadian and U.S. telecommunications industry. Although change was produced slowly, over a period of more than thirty years, the nature and extent of change was monumental. It was also improbable given the size of the monopolists (both in terms of revenues and numbers of unionized employees), the high quality and reliability of the service they were providing, and their contribution to national defense through the research they performed for the military. What were the differences between the U.S. and Canadian approaches to deregulation? How do we explain these differences? And what do these differences tell us about the process through which policy change has been effected in each country?

As noted in the introduction to Part II, the story of U.S. deregulation is essentially the story of the breakup of AT&T's vertically integrated monopoly. AT&T's competitors, such as the Carterfone Company, manufacturers of microwave equipment, and MCI, played important roles in spurring change. Other private sector actors included corporate users of AT&T services, such as the major broadcasting networks, who were dissatisfied with AT&T, and corporately funded policy institutes who through their studies and reports legitimized liberalization. In various ways, and according to their individual self-interest, these private sector actors, including AT&T, attempted to influence the nature and the pace of change. But the reins of power were in the hands of public sector actors, particularly at the federal level: the Federal Communications Commission (FCC), the Justice Department, the courts, and the executive branch. What makes U.S. deregulation distinct from the Canadian experience, and other national experiences, is the role that the Justice Department, under the auspices of antitrust law, has played in shaping industry structure. The Justice Department initiated antitrust action against AT&T three times during the twentieth century. The second suit produced the Consent Decree of 1956; the third suit produced the AT&T divestiture.

The Consent Decree acted as an absolute barrier to AT&T diversification into non-common carrier markets. Moreover, it discouraged the computer industry from participating in markets for hybrid services. It was the 1956 Consent Decree that set the stage for, indeed necessitated, the *Computer I, Computer II,* and *Computer III* rulings. For all intents and purposes the decree was the cornerstone of the policy of non-convergence and was a powerful sanction for the policy that supported and protected AT&T's monopoly. The third suit, which resulted in the Modification of Final Judgement (MFJ), produced another fundamental change in industry structure. By opening the door to AT&T diversification, the MFJ set in motion the process of industry convergence, which, by the end of the 1990s, would culminate in a policy that was more tolerant of mergers and alliances between the entertainment, telecommunications, and cable industries. It would be difficult to understate the importance of antitrust law and, in particular, the Justice Department in shaping the course of U.S. deregulation.

The other singular characteristic of U.S. deregulation has been the central role that the courts have played in making policy: the *Hush-A-Phone* decision, the *Execunet* decisions, and Judge Harold Greene's oversight of the MFJ. It was the *Hush-A-Phone* decision that set the stage for the *Carterfone* decision and the subsequent opening of the terminal equipment market. The *Execunet* decisions effectively liberalized the long distance market. These were landmark events that effectively changed the course of telecommunications policy in the United States. Judge Greene's intervention in the MFJ transformed his court into one of the principal arenas for ad hoc policy making from 1982 until the passage of the Telecommunications Act of 1996. Conversely, the courts have also played an important role in forestalling change through the appeals process.

At first the FCC was reluctant to sanction competition in any of the markets controlled by AT&T. To the extent that the FCC was prepared to liberalize at all, it was on the periphery of markets controlled by AT&T. But these attempts were superseded by the courts, that is, the *Hush-A-Phone* and *Execunet* decisions. It was not until the *Computer III* ruling (1985) that the FCC embraced competition with its open network architecture policy. But the stage had already been set for this reversal of policy by the MFJ (1982) and by the opening of the long distance and terminal equipment markets, both of which were brought on by the courts. Moreover, the open skies policy that governed satellite communications was an initiative of the executive branch. In short, the FCC was a reluctant participant in liberalization. When it finally embraced competition it would do so under the guise of a policy of pro-competitive regulation not laissez-faire.

The process of telecommunications liberalization in Canada would differ from that of the United States in a number of important respects. We will begin our analysis by reviewing the principal actors. The federal cabinet, the Department of Communications (now part of Industry Canada), the Canadian Radio-television and Telecommunications Commission (CRTC), the Restrictive Trade Practices Commission (RTPC) and its Director of Investigation and

Research, the provincial governments, and the courts were the principal public sector actors. The principal private sector actors were the following: the incumbent common carriers (including their long distance consortium Telecom Canada), their competitors (including the independent telephone companies and resellers), and competitive equipment suppliers. As was the case in the United States, the public sector actors held the reins of power.

It is important to remember that until the 1970s Canada's policy regime was defined not only by the monopoly mosaic but also by the policy of using telecommunications to advance cultural, national sovereignty, and national unity goals. This began to change in 1976 when the authority to regulate telecommunications under the Railway Act was transferred from the Canadian Transport Commission to the CRTC. This was primarily an organizational change. It did not entail a legislative change with respect to the obligations of the common carriers, nor did it entail a change in the CRTC's mandate with respect to the manner in which the telecommunications industry should be regulated. Notwithstanding this fact, as early as 1977 the CRTC was beginning to nudge the telecommunications industry in the direction of greater competition. This occurred when the commission opposed Telesat participation in the Trans Canada Telephone System (TCTS). However, the federal cabinet did not share the CRTC's enthusiasm for intermodal competition. It was from this point on, however, that the CRTC could be identified as a force for change, albeit managed, incremental change.

When the CRTC took these first steps in the direction of a more competitive industry structure, it was not acting in a void. U.S. deregulation would have a significant demonstration effect on Canadian policy making. When deregulation lowered the price of long distance telecommunications in the U.S. domestic market, and produced more variety in the U.S. terminal equipment market, it was not long before business users in Canada took notice and began to lobby the CRTC for change. The CRTC was receptive to the needs of Canadian business while being reluctant to move precipitously. What emerged was a policy process dominated by the case-by-case adjudicative approach orchestrated primarily by the CRTC.

It was at this point that a critical question arose: How could the CRTC reconcile competition with the public service obligations of the common carriers? Or more precisely, how could it balance the interests of business and residential users of telecommunications services? This problem would preoccupy CRTC policy making for almost two decades. In 1979, *CNCP Telecommunications, Interconnection with Bell* (TD CRTC 79-11), and again in 1985, *Interexchange Competition and Related Issues* (TD CRTC 85-19), the commission drew a line separating competitive from non-competitive sectors of the market based on the principle that public voice service (Message Toll Service [MTS] and Wide Area Telephone Service [WATS]) should continue to be provided on a monopoly basis. In both decisions the CRTC invoked the public service obligations of the common carriers, and more specifically, their

ability to subsidize local service using revenues derived from the other parts
of their business, as a key to preserving the public policy goal of universality.

In 1992 (TD CRTC 92-12) the commission reversed itself. The policy reversal was notable in at least two respects. First, very little had changed with
respect to the problematic, which opposed competition and universality between the years 1985 (and indeed 1979) and 1992. The regulatory options
that were available to the CRTC in the 1990s were essentially the same as
those that were available to the commission in the 1970s and 1980s. Moreover, the record of U.S. deregulation was inconclusive. Second, once the
CRTC sanctioned competition, it went further in its efforts to liberalize the
market for MTS and WATS than almost any of the participants in the proceeding had anticipated, including Unitel.

Although policy was created by the CRTC through an essentially ad hoc adjudicative process, one would be remiss if one did not acknowledge that this approach was frequently punctuated by public consultation. These consultations
were orchestrated primarily by the Department of Communications and resulted in a number of government reports and statements of policy. This was particularly true during the period that preceded vigorous liberalization (from 1969
to 1979) when three reports were produced: the Telecommission report (Department of Communications 1971a), the Green Paper (Department of Communications 1973) and the Clyne Committee report (Department of Communications 1979). While these reports did not dictate policy to the CRTC, they
set the parameters of the public debate on telecommunications, gradually moving policy away from the old regime that viewed telecommunications as a tool to
advance other political and cultural objectives.

In Canada the courts did not play a major role in setting communications
policy once jurisdiction had been transferred to the CRTC. There were no reversals of CRTC policy comparable to the *Execunet* decision or the decision
of Judge Harold Greene to oversee the MFJ. When the court did intervene, it
did so to reverse the policies of the Canadian Transport Commission, the
CRTC's predecessor, with respect to telecommunications. But the Supreme
Court did play a key role in resolving the jurisdictional issue of federal vs.
provincial authority to regulate Canada's provincial telephone companies.
When the court ruled in the *AGT* case (1989), it resolved the long-standing
problem of how to create a national policy for telecommunications, a preoccupation of Canadian policy makers. In *La Régie des Services Publics v. Dionne*
(1978), the Supreme Court recognized Ottawa's jurisdiction to regulate the
cable industry. To summarize, the Canadian courts were not zealous interveners in the policy process, nor did public and private sector actors view the
court as a legitimate arena for the resolution of disputes. But when the courts
were called upon to resolve fundamental constitutional issues concerning jurisdiction, their rulings in favor of federal jurisdiction would have far-reaching
consequences for the future of the industry.

It is extremely significant that Canadian liberalization did not result in the dismantling of the vertically integrated business of Bell Canada or any other telephone company. In Canada there was no sustained antitrust action designed to separate the three pillars of Bell's integrated business: terminal equipment, long distance, and local service. At one point, the Director of Investigation and Research recommended that Bell should divest itself of its terminal equipment business, but the recommendation was not even given serious consideration by the RTPC. The Canadian model removed the regulatory protections that had shielded Bell Canada from competition, while allowing Bell to retain its vertically integrated business. In other words, Bell lost its monopoly, but it did not lose its vertically integrated business.

The American and Canadian approaches to liberalization were based on very different assumptions concerning the ability of a regulatory agency to control the anti-competitive behavior of dominant firms with market power.[1] The American model was based on the assumption that regulation could not be expected to effectively control the anti-competitive behavior of a firm with market power. Conversely, the Canadian model gave little credence to the argument that Bell was a firm with market power. More importantly, the Canadian model was grounded in the confident assumption that the regulator and the regulated could work together to ensure that competition was fair. This would be accomplished through the exchange of accounting information, based on accounting models that had been established by the regulatory agency after consultation with the regulated telephone companies. At no point did the federal cabinet, the Department of Communications, the provincial governments of Ontario and Quebec, or the CRTC seriously consider an approach that would have seen Bell Canada reduced and weakened through the divestiture of its vertical operations. On the contrary, the federal government allowed Bell Canada to strengthen its competitive position when it did not oppose the Bell Canada reorganization of 1987.

How does one explain this fundamental difference in approach? In Canada the economic dogma was superseded by geopolitical considerations. Although Bell Canada is a large company in the context of the Canadian market, the Canadian market is itself small in relation to that of the United States. Bell may have fought hard to maintain its regulatory protections from competitive entry, but it did not have to lobby nearly as hard keep its business intact. Although never articulated as a matter of explicit policy, it was a matter of implicit policy that Canada wanted to participate in emerging liberalized Canadian, U.S., and international telecommunications markets. By default, Bell Canada would become an important instrument of this policy.

To summarize, the deregulation of the Canadian and U.S. telecommunications industries has differed in a number of important respects. Canadian deregulation generally followed U.S. deregulation by at least ten years. In many cases this allowed for a more orderly transition to competition. The courts in the United States played an important role in forcing change (i.e., *Execunet*). This was never the case in Canada. However, in Canada the courts played an important role in resolving in a fairly definitive manner the problem of jurisdiction. In the United States the courts continue to play a role in settling jurisdictional issues, but this is part of a continuous process. Two-tiered responsibility still exists with the courts in the middle assuming the role of arbiter of case-by-case disputes involving the different levels of government. Although the parliamentary system allows for more direct political control of the regulatory agency, the CRTC has been relatively autonomous. Moreover, the courts have not been very active in reversing CRTC policy. In the United States the executive branch has placed its stamp on the FCC through its authority to name commissioners, and in particular, the FCC chairperson. Although the U.S. system does not allow for executive branch reversal of FCC decisions, the courts have played an important role in setting de facto policy. This has diminished the autonomy of the regulatory agency, the FCC. In the United States, antitrust law has played a key role in determining industry structure and the Justice Department has played an important role in shaping policy. In Canada, antitrust law has played a negligible role in shaping policy for the telecommunications sector. This has been the case primarily because there has been little political will to interfere in the corporate structure of Bell Canada. Finally, the Canadian system is generally more confident of the ability of the regulator to control the anti-competitive behavior of firms with vertically integrated operations such as Bell Canada. In the United States there has been a greater tendency to impose structure separation as a solution to the problem of vertical concentration.

NOTE

1. The term "market power" refers to the ability of a firm to control prices in a given market. More specifically it refers to the ability of a firm to set prices at a level significantly above the level of its marginal costs and for a sustained period of time. A firm with market power may face competition, but its share of the market is so great that its competitor's lower prices have little impact on its pricing decisions. According to economic theory, market power may result from a number of factors: the control of essential patents, the existence of important economies of scale, and the control of essential inputs to its core business (i.e., vertical integration).

General Conclusion

In chapter three we identified three periods of economic regulation: Progressive, New Deal, and Great Society. From this analysis we concluded that economic regulation has been introduced as a response to varied economic conditions. There was no single logic or rationale for economic regulation. Economic regulation and the institutions it produced were the product of at least three historically distinct types of economic problems: excessive competition, market collapse, and externalities. This analysis was used to underscore the fact that the impetus behind economic regulation has not been solely the protection of consumers. At various times economic regulation has served the interests of producers as well. But our analysis stopped short of describing the logic and the specificity of the current period. We will now consider the deregulation of Canadian and U.S. telecommunications in the context of larger, more pervasive changes in state-economy relations.

In *Regulatory Politics in Transition* (1993), Marc Allen Eisner has coined the term "regulatory regime" to designate the three periods of economic regulation mentioned earlier. Eisner defines a regulatory regime as "a historically specific configuration of policies and institutions which structures the relationship between social interests, the state, and economic actors in multiple sectors of the economy" (1). According to Eisner, a new regulatory regime emerges as the result of a major shift in state-economy relations encompassing broad shifts in policy and new institutional arrangements. The term is not meant to designate shifts in policy or institutional change within a single sector, such as telecommunications, but rather major historical change in state-economy relations affecting numerous sectors of the economy. Although Eisner substitutes the terms, market regime, associational regime, and societal regime for the terms, Progressive, New Deal, and Great Society, his analysis of the periods essentially corresponds to that of Robert Britt Horwitz (1989).

271

Eisner has also coined the term "efficiency regime" to describe the contemporary period. According to Eisner the efficiency regime is the product of "economic stagflation and growing foreign competition in the 1970s and 1980s" (1993, 9). The primary goal of the efficiency regime is the "elimination of policies that interfere with market mechanisms or impose large compliance costs" (9). The principal policy initiatives of the regime in the United States included: (1) directives from the executive branch of government mandating formal economic analysis of regulation in a number of industry sectors (i.e., cost-benefit analysis of regulation); and (2) the passage of deregulatory legislation, initially in the airline, surface transport, and finance industries. The first is characteristic of efforts at regulatory reform; the second is synonymous with outright deregulation.

Eisner's analysis highlights two defining characteristics of the efficiency regime: (1) the ascendency of economic analysis and the profession of economics in policy making; and (2) the fact that popular awareness of the regulatory problem was not widespread, and that public support for reform or deregulation was non-existent. He states, "a search for the democratic roots of the efficiency regime is fruitless" (1993, 173). Eisner attributes the emergence of the efficiency regime to three factors: (1) a change in the interest group system, (2) presidential initiatives aimed at the reform of regulation, and (3) the introduction of new administrative practices, including changes in staffing policy at the regulatory agencies (177). He states:

> Corporations responded to the combination of new social regulations and deteriorating economic conditions by demanding regulatory relief. The demands of corporations and trade associations were clearly focused on regulation, whereas popular concerns focused overwhelmingly on inflation and stagnant economic growth. However, these popular concerns were linked with prevailing economic critiques of regulation and with the argument that regulatory administration could be reformed through the application of economic analysis in regulatory decision-making. These doctrines found a clear expression in a series of executive orders. The executive review process, in turn, forced institutional changes, thus creating a bureaucratic constituency for deregulation and regulatory reform. Although it is difficult to determine whether corporate PAC contributions were decisive, such funding reinforced the political expediency of deregulation and may have been instrumental in increasing congressional support for deregulation. (178)

Eisner's analysis of the efficiency regime confirms the analysis presented in the previous chapters that the deregulation of the telecommunications sector was primarily a response to the demands of corporate users who viewed competition as a lever to reduce telecommunications costs, and in particular costs related to long distance communications. But it adds another dimension, which the preceding chapters only explored summarily. The new regime was not only the product of the ascendency of a corporately funded, economic critique of regulation from outside the regulatory agency—expert testimony, think tanks, and policy institutes. It was also a product of change from within the regulatory agencies as staffing decisions were made to replace political scientists and lawyers with economists. Eisner states, "as economists gained a greater role within the agencies, their hostility toward many of the policies they were charged with implementing became evident, and they promoted deregulatory experimentation, especially when such experimentation was supported by sympathetic agency executives" (1993, 194).

What emerges is a coherent and fairly comprehensive portrait of the new regime and the forces that produced it. To summarize, the efficiency regime is characterized by the following:

1. The ascendancy of economic methods in the analysis of regulatory policy and practices
2. The ascendancy of the profession of economics within the regulatory agencies
3. Rising costs associated with Great Society (societal regime) produced a corporate backlash, whose initial objective was to reduce the costs of regulation, but which ultimately crystallized into a general critique of economic regulation
4. The linking in the minds of the public of inflation and economic stagnation with the problem of regulation
5. The politicization of regulatory reform at the level of the executive branch of government that saw several presidencies (Ford, Carter, and Reagan) promote regulatory reform, based on economic criteria, as a way of improving the efficiency of regulation and lowering its costs to business
6. The introduction of deregulatory legislation in certain sectors

The concept of regulatory regimes, as presented by Eisner, applies to significant shifts in policy and institutions that are historical in nature and macrosectoral in their reach. As such, the concept enables us to situate sector-specific change within a larger context of historical change in state-economy relations. On the other hand, it is important to remember that these changes are manifest in specific sectors and are subject to sector-specific constraints. For example, in the case of telecommunications, a description of the new regulatory regime would have to take into account whether there was change or continuity with respect to the problem of jurisdiction: federal vs. state (or provincial) authority. This is a sector-specific characteristic that may have little or no resonance in other industries. In what follows we will adapt the concept of regulatory regime to the analysis of historical shifts in policy and institutions at the level of the telecommunications industry.

What are the defining characteristics of the new regulatory regime at the level of the telecommunications sector? The analysis presented in the previous chapters suggests that the new regime in both the United States and Canada exhibits the following characteristics. It is:

1. Tolerant of vertical and horizontal concentration—abuses of market power are matters for antitrust law not economic regulation (antitrust solutions to market power)
2. Opposed to service segmentation based on industry segregation (convergence)
3. Committed to pro-competitive regulation—as opposed to traditional monopoly regulation (pro-competitive regulation)
4. Favorable to the use of regulatory forbearance, albeit on a case-by-case basis (discretionary regulatory forbearance)
5. Grounded in the principle of the coordinated use of public policy and policy instruments at all levels in order to achieve broad national policy goals (interrelatedness of policy)
6. Open to the preservation of traditional public service values—universality, equitable access, quality of service, rural and regional development, and support for disadvantaged groups—and the tools traditionally associated with their promotion—system-wide price averaging and the use of direct and indirect subsidies (the new universality)
7. Committed to the use of technology policy in traditional sectors of public control (health and education) as a lever for the development of information and communications industries (leveraging the public sector)

In light of the analysis presented in chapters five to eight, the first four characteristics of the new regulatory regime are largely self-explanatory: reliance on antitrust, industry convergence, pro-competitive regulation, and discretionary regulatory forbearance.

Probably the most singular example of the fifth characteristic of the new regime (the interrelatedness of policy) is the link between the liberalization of the local exchange market, convergence policy (which allows common carriers to become information providers and video programmers), and the national strategy to promote the information superhighway. By opening the local services market to new competitors and by allowing incumbent carriers to participate in new markets for programming services, policy makers hope to create conditions that will favor investment in broadband infrastructure. In theory, facilities providers will be able to reap the benefits of investment in infrastructure at the level of both content and carriage. None of these policies stands in isolation; they are organically linked. Similarly, the new more tolerant approach to economic concentration will encourage mergers and alliances between the entertainment, telephone, and cable industries. It is hoped that these vertically integrated conglomerates will, as the result of their market power and greater access to capital, have the critical mass necessary to invest in broadband facilities to the home. It is not our intention to suggest that these policies are motivated by the singular goal of promoting investment in network infrastructure. Ostensibly the policies concerning competition, convergence, and concentration are designed to promote innovation, greater efficiency, and greater choice of services and products for the consumer. But they are also designed to effect changes in the industry structure in the direction of greater concentration of capital, changes that policy makers believe will be conducive to investment in networks. We have referred to this policy dimension as the principle of interrelatedness. In the example cited earlier, it concerns the interrelatedness of infrastructure policy and policies concerning content and carriage, industry convergence, and economic concentration. However, it could just as well apply to the interrelatedness of policies for the broadcasting and telecommunications industries, or within the telecommunications industry between policy frameworks governing local service, long distance, and equipment manufacture.

For the most part, the new regime is committed to the public service goals that were the cornerstone of the monopoly regime for telephony. In fact there is a more explicit statement of these goals in both the Canadian and U.S. telecommunications acts. Recall that the Canadian Telecommunications Act establishes the following objective for telecommunications policy, "to render reliable and affordable telecommunications services of high quality accessible to Canadians in both urban and rural areas in all regions of Canada" (7.(b)). The U.S. Telecommunications Act is even more explicit. Section 254, entitled "Universal Service," mandates the creation of a joint state-federal board with responsibility for implementing a universal service policy (*see* chapter six, section six, for a listing of the principles that are to guide the board.) The U.S. Telecommunications Act innovates in a number of important respects: it creates a universal service fund, administered by the joint board, to which all telecommunications service providers must contribute; the subsidies distributed from the fund are direct and transparent; the subsidies are targeted at specific constituencies; and the act embraces a flexible definition of what services should be included as part of a universal service policy. In theory, today's advanced services could one day be considered basic.

Although there is a more explicit commitment to universal service in the new regime (both Canadian and American), the success of the policy will ultimately depend on how this policy is implemented. The Canadian statute essentially leaves it to the Canadian Radio-television and Telecommunications Commission to monitor the price and availability of services, using its authority to review and set rates for Canada's major carriers as its principal policy tool. The Canadian act does not prescribe that subsidies be transparent. Although the U.S. legislation embodies a more explicit and expansive commitment to universality, the fact that the mechanisms for generating and distributing the subsidy in the United States are transparent guarantees that the policy will become highly politicized. Regardless of how successful regulators are in implementing the policy, it is extremely significant from the point of the new regulatory regime that the commitment to universality has been carried over from the old regime. The policy of universality, now that it is enshrined in legislation, will act to counterbalance the other provisions of the telecommunications acts that favor markets and competition as the ultimate regulators of service availability, quality, and price.

During the Cold War, the defense industry was the focal point for research and development of advanced information and communications technologies. Although the defense sector continues to play a major role in the development of these technologies, the winding down of the defense sector has meant that government, particularly in the United States, has sought other channels to promote the development of advanced information and communications technology. The principal "beneficiaries" of this policy have been the health and education sectors. This opening is clearly manifest in the U.S. Telecommunications Act, which prescribes that the universal service fund be used to subsidize access to broadband networks in the health and education sectors. In both Canada and the United States, governments at both the national and state/provincial levels have established programs to subsidize access to broadband networks and the Internet from schools and health institutions. In many instances it would appear that these programs have more to do with an industrial strategy for the computer and communications industries than they do with meeting the real needs of the educational and health care systems.

The seven dimensions are common to the new regulatory policy regimes of both the United States and Canada. To summarize, they include the following: reliance on antitrust law to control abuses of market power, industry convergence, pro-competitive regulation, discretionary regulatory forbearance, interrelatedness of policies, the new universality, and leveraging the public sector to facilitate the transfer of advanced communication/information technologies. This is not meant to suggest that the application of these principles will lead to identical or even comparable results. As we have suggested throughout this work, Canadian and U.S. institutions, industry structures, and traditions are sufficiently different to ensure diverse results. However, as guiding principles they appear to be structuring policy in the new regimes in both countries.

On the other hand, the Canadian and U.S. regimes differ in at least two important respects: (1) how their regimes deal with the problem of vertical concentration; and (2) the extent to which the regimes have been able to resolve the problem of split jurisdiction to regulate (or deregulate) the telecommunications sector.

Although the new regime is generally more tolerant of economic concentration, including the vertical integration of operations, policy makers still have well-founded reservations about the potential for abuse when common carriers provide everything from content, to equipment, to multiple telecommunications services in a competitive environment. The Canadian approach to this problem has been to allow the regional telephone companies, such as Bell Canada, to provide telecommunications facilities and services as part of an integrated operation subject to accounting controls. The U.S. approach has been to first dismantle vertical operations (the AT&T divestiture), and more recently to open the door to the integrated provision of local and long distance service. But this is subject to the Bell Operating Companies (BOCs) compliance with a competitive checklist and the presence of a competitor for interconnected service (*see* chapter six, section six). When these conditions are met, a BOC may provide long distance service through a structurally separate subsidiary. After three years a BOC may apply to the Federal Communications Commission (FCC) for relief from the structural separation requirement. The Canadian and U.S. approaches to vertical integration are clearly different. In practical terms these differences mean that Canadian firms are currently operating under conditions about which their U.S. counterparts can only dream. At a more basic level the different approaches suggest that the U.S. regime is characterized by a more healthy respect for dominant firms and a more deep-rooted scepticism of their willingness to compete fairly. Conversely, one might say that the Canadian regime is grounded in a more confident view of regulation, and more specifically, of the regulator's ability to monitor the behavior of dominant firms as a way of controlling abuses. To summarize, the U.S. regime views the vertically integrated provision of telecommunications products and services as a potential threat to fair competition and favors structurally separate subsidiaries as the most effective remedy; the Canadian regime plays down the threat and views accounting controls as an effective remedy.

Finally, the regimes differ with respect to the degree to which jurisdictional conflicts play a role in shaping policy. Essentially, the Canadian regime resolved the problem of federal jurisdiction to regulate (deregulate) telecommunications prior to the passage of the Canadian Telecommunications Act. The act consolidated these changes in favor of federal jurisdiction. Conversely, since the passage of the U.S. Telecommunications Act in 1996, conflicts between local jurisdictions and the FCC have heated up. A recent and symptomatic example of this conflict concerns an attempt by the city of Portland, Oregon, to force AT&T to open its cable network to unaffiliated Internet service providers. Anticipating the arrival of the BOCs in the long distance market, AT&T is attempting to gain control of its own local network facilities through the purchase of cable systems. To this end AT&T has acquired Tele-Communications, Incorporated, and has announced a deal to purchase the Mediaone Group. However, both of these cable companies currently provide Internet access via cable modems through a network architecture provided by @Home Network. (Essentially, @Home Network provides the Internet backbone network and cable modems; the cable operators provide the local facilities.) Because the cable operators have an exclusive contract with @Home Network, it is impossible for other information service providers (ISPs), such as America On Line (AOL), to offer Internet service using the cable network. AT&T appealed to the courts for relief from this requirement. In a decision rendered June 3, 1999, AT&T's request was rejected by the district court (*AT&T Corp. Tele-Communications, Inc.; TCI Cablevision of Oregon Inc., and TCI of Southern Washington v. City of Portland and Multnomah County*, United States District Court for the District of Oregon. June 3, 1999.) The decision is under appeal and the FCC has filed a friend of the court brief arguing that it is a matter for the FCC, and not the city of Portland, to decide who should have access to high-speed cable lines. It is important to note that Portland is not the only jurisdiction attempting to shape Internet policy; moreover, it could take years before the case is finally resolved through the appeals process. The contrast with Canadian treatment of this issue is striking.

On September 14, 1999, the CRTC ordered cable carriers who provide Internet service to open their networks to competing Internet service providers. These operators are required to make their networks available for resale at a discount of 25 percent from their lowest retail Internet service rate (TD CRTC 99-11). The cable companies have not challenged the decision in the courts because there is little basis in the Canadian system for such a challenge. It is interesting to note that the position that has been taken by the CRTC is essentially that of the city of Portland and the competitive ISPs.

It is extremely significant that the principles of the new regime have been enshrined in major legislation in both the United States and Canada. In effect, the passage of the U.S. and Canadian telecommunications acts marks the end of the old regime, both symbolically and in terms of actual policy practices. But it does not mark the beginning of a new era of state-industry relations in the telecommunications sector, since this has been underway for over two decades. In practical terms, the acts mark the culmination of a long process through which the terminal equipment, long distance, and enhanced services markets were liberalized. It also marks the beginning of a new era and a new policy framework designed to bring competition to the last infrastructure bottleneck, the local exchange, and to extend the availability of enhanced services via broadband facilities to the residential market, the new universality.

What are the limitations and contradictions of the new regime and what impact will these have on the future direction of policy and the availability of services? It is clear that there will be no significant reversals or changes in policy orientation at the level of the telecommunications policy regime in the near future. This becomes evident when the policy regime for telecommunications is situated in the larger historical context of the efficiency regime. There is little evidence to suggest that the regime has run its course or that another regime is around the corner. On the other hand, one may hypothesize that the limitations of the current policy regime for telecommunications may be traced to the social, political, and economic forces that produced the efficiency regime at a macro level. Ostensibly, the regime was a response to runaway inflation coupled with stagnant economic growth. A new monetary policy coupled with a decade of economic growth has effectively removed these conditions. To the extent that these problems have been resolved, whether as a result of the efficiency regime or through other changes in economic policy, one can conclude that the regime has lost its urgency. In many respects it may be considered yesterday's solution to yesterday's problem. Stated in other terms, it is no longer possible or even necessary to link the policies associated with the efficiency regime with the problem of stagflation. As a result, we have seen that these policies have found a new justification. At the level of the telecommunications industry, policies associated with the regime are now proposed as solutions to the problem of high service costs and slow rates of innovation associated with infrastructure bottlenecks and monopolistic industry structures.

The other elemental component of the efficiency regime was the centrality of economic models and economic theory to policy making. Neo-classical economic theory has fairly dominated perceptions of both the objectives of policy and the means required to obtain them. Policies of regulatory forbearance, procompetitive regulation, and even convergence and tolerance of economic concentration are all grounded in the assumption that vigorous competition will establish itself and will be sustainable. In this manner the regime has created not only its own agenda for policy, but also its own standards for evaluating the success or failure of policy. Whether measured in terms of industry structure (the number and organization of firms competing to provide services) or in terms of firm behavior (competitive price setting, innovation, diversity, and quality of services) competition is now the standard for measuring success.

The stage will soon be set for a major reassessment of the new policy regime for telecommunications. This will occur in the short to medium term as the competitive regime, which has been consecrated by the passage of the telecommunications acts, confronts the limits of its ability to achieve its own stated objectives. There is already evidence to suggest that the new regime is failing in two respects: (1) the refusal of firms to compete in markets for network facilities; and (2) the emergence of a "digital divide" separating business and affluent residential users from the rest of the population. In their first triennial review of the U.S. Telecommunications Act, the Consumer Federation of America and the Consumers Union argue that measures should be taken to protect consumers against the failure to compete. The authors note:

- Instead of becoming vigorously competitive, the telecommunications and cable industries have become highly concentrated
- Instead of significant declines in prices, we have sharp increases in cable (21%) and in-state long distance (10%), and stagnation in local phone and interstate long distance rates

- Instead of rapid deployment of advanced technologies from increased private sector investment, we have a growing "Digital Divide" between those who make intensive use of the telecommunications network and those who do not (Cooper and Kimmelman 1999, iv)

The authors charge that firms have responded to the opportunities created by the Telecommunications Act by implementing "business models that focus on a narrow, premier segment of the market" (vii). Rather than developing markets catering to the needs of average consumers, the industry has opted to market services to high-end, intensive users of communication services.

It would appear that the new regime has entered a very difficult and critical phase, one that will test the validity of some of its most fundamental policy principles. At one level the litmus test for the new regime will be the extent to which broadband networks will be available and affordable. But just as important will be the extent to which access to services is based on open models. Network operators are once again withholding investment in facilities on the grounds that an open model, which "relegates" to them the status of a simple common carrier and affords them no privileges at the service level, is uneconomic. While the telephone companies seek relief from the unbundling requirements of the Telecommunications Act, AT&T is attempting to regain control of the local access market through the acquisition of the nation's largest cable franchises and the application of the cable industry's traditional closed access model to the Internet. The reconfigured wire-line industries are on a collision course with regulators that will test the efficacy of the new regime, and the neo-classical economic theory that underscores it, to a degree that was impossible during the previous cycles of liberalization and deregulation. Moreover, it is likely that there will be stiff public resistance to attempts to impose a closed access model to the Internet. This will occur because there is already a large installed base of network users who are familiar with the open dial-up model for Internet access and relatively familiar with the role that an open approach to network architecture has played in the phenomenal growth of the Internet. The demonstration effect of the Internet success story will act as a substantial counterweight to pressures for closed systems. As the industry becomes more concentrated with facilities operators refusing to compete, while at the same time failing to invest and attempting to impose closed system models of access to the Internet, the fundamental precepts of the new competitive regime will be severely tested.

In the United States, the pressure will be manifest at the interstices between the regulatory agencies and the courts and between local/state and federal jurisdictions. In Canada, the shortcomings of policy will become evident as the cable and telephone industries exert pressure on the CRTC for a reversal of policy, potentially with support of government. The most probable result in both countries will be a policy stalemate, a period of "muddling through" that will see the competitive regime lose much of its legitimacy, but without the possibility of reverting back to the old regime of structural separation and diligent oversight of rates. As long as the efficiency regime prevails at a macro level, it is hard to imagine a serious challenge to the new model—based as it is on neo-classical economics of the Chicago school and enshrined in the telecommunications acts. It is difficult to imagine how public policy for the telecommunications sector will be able to escape its enslavement to neo-classical dogma. In fact, policy for the telecommunications industry will surely be caught in a conceptual no-man's-land between the alleged promise of neo-classical economics and the hard reality of the facilities providers' near categoric refusal to compete with one another, all the while eliminating competition whenever it emerges in the margins of their industry. Moreover, as we have demonstrated, this pattern of corporate behavior has been one of the most consistent features of the industry for over a century. No doubt the policy debate will remain stale and sterile until a new enlightened economics of network industries emerges from the shadow of neo-classical dogma, an economics that takes into account historical patterns of corporate behavior in the industry, and breaks away from the premise that these markets are essentially like any other. The principal irony of pro-competitive regulation is that it exists at all. This is not so much a failure or contradiction of policy or of regulation as it is a failure of the economics that informs policy.

Bibliography

AUTHORED WORKS

Akwule, Raymond. *Global Telecommunications: The Technology, Administration, and Policies.* Stoneham, Mass.: Focal Press, 1992.

Anderson, James E. "Economic Regulation." In *Encyclopedia of Policy Studies,* ed. Stuart S. Nagel. New York: Marcel Dekker, 1983.

Arnst, Catherine, Garry McWilliams, and Amy Barret. "The Return of Ma Bell?" *Business Week,* June 9, 1997: 30–32.

Association des compagnies de téléphone du Québec. *Mémoire présenté au CRTC en réponse à l'avis public télécom 1990–73, 3 août 1990,* 1991.

Aufderheide, Patricia. *Communications Policy and the Public Interest: The Telecommunications Act of 1996.* New York: The Guilford Press, 1999.

Averch, Harvey, and Leland L. Johnson. "Behaviour of the Firm under Regulatory Constraint." *American Economic Review* 52 (December 1962): 1052–1069.

Babe, Robert E. "Predatory Pricing and Foreclosure in Canadian Telecommunications." *Telecommunications Policy* 9, no. 6 (1985): 329–333.

————. *Telecommunications in Canada.* Toronto: University of Toronto Press, 1990.

Bannock, Graham, R. E. Baxter, and Evan Davis. *Dictionary of Economics.* London: Hutchinson Books in association with The Economist Books, 1989.

Baumol, William J., John C. Panzar, and Robert D. Willig. *Contestable Markets and the Theory of Industry Structure.* New York: Harcourt Brace Jovanovich, 1982.

Bell Canada. *Guide Préliminaire en vue de devenir fournisseur de service sur ALEX,* 1988.

Bell Canada Enterprises. "Corporate Organization." BCE Web Page. (http://bce.ca/bce/fs/e/about/corporate/organization). Consulted: October 21, 1997.

Bercuson, David J., and J. L. Granatstein. *The Collins Dictionary of Canadian History: 1867 to the Present.* Don Mills, Ontario: Collins, 1988.

Berman, Paul J. "Computer or Communications? Allocation of Functions and the Role of the Federal Communications Commission." In *High and Low Politics, Information Resources for the 1980s,* ed. Anthony Oettinger. Cambridge, Mass.: Ballinger, 1977.

Bernstein, Marver H. *Regulating Business by Independent Commission.* Princeton, N.J.: Princeton University Press, 1955.

Bethesda Research Institute. *Separate Subsidiaries and Structural Separation in United States Telecommunications: Conceptual Analysis, Applications, and Prognosis.* Prepared for the Ministry of Transportation and Communications, Province of Ontario by Research Studies Division, Bethesda Research Institute, Ltd. Bethesda, Maryland, 1985.

Birrien, Jean-Yvon. *Histoire de l'informatique.* Paris: PUF (Que Sais-Je No 2510), 1990.

Bolter, Walter, James W. McConnaughey, and Fred J. Kelsey. *Telecommunications Policy for the 1990s and Beyond.* New York: Sharpe, 1990.

Brenner, Daniel L. *Law and Regulation of Common Carriers in the Communications Industry.* Boulder, Colo.: Westview, 1992.

Breyer, Stephen. "Analyzing Regulatory Failure: Mismatches, Less Restrictive Alternatives, and Reform." *Harvard Law Review* 92 (January 1979): 547–609.

Brock, Gerald W. *The Telecommunications Industry: The Dynamics of Market Structure.* Cambridge, Mass.: Harvard University Press, 1981.

———. *Telecommunication Policy for the Information Age: From Monopoly to Competition.* Cambridge, Mass.: Harvard University Press, 1994.

Brotman, Stuart. *The Telecommunications Deregulation Sourcebook.* Boston: Artech House, 1987.

Bruce, Robert R., Jeffrey P. Cunard, and Mark D. Director. *From Telecommunications to Electronic Services.* Washington: Debevoise and Plimpton, 1986.

Canada. Canadian Radio-television and Telecommunications Commission. *Annual Report, 1981–1982.* Ottawa: Minister of Supply and Services, 1982.

———. *Annual Report, 1990–1991.* Ottawa: Minister of Supply and Services, 1991.

———. *Competition and Culture on Canada's Information Highway: Managing the Realities of Transition.* Ottawa: Public Works and Government Services Canada, May 15, 1995.

———. Fact Sheet, "Price Cap Regulation." May 15, 1997.

Canada. Industry Canada. *Connection, Community, Content. The Challenge of the Information Highway.* Final report of the Information Highway Advisory Council. Ottawa: Minister of Supply and Services, September 1995a.

———. *Notice no. DGTP-008-95 (July 1995) Review of Candian Overseas Telecommunications and Specifically Teleglobe Canada's Role.* Ottawa: Industry Canada, 1995b.

———. *Convergence Policy Statement.* Ottawa: Industry Canada, July 10, 1996.

———. News Release. "Canada Welcomes Telecom Deal." February 15, 1997.

Canada. Federal-Provincial Conference of Communications Ministers. *Federal-Provincial-Territorial Memorandum of Understanding* (Edmonton Accord), signed on April 3, 1987.

Canada. Restrictive Trade Practices Commission. *Telecommunications in Canada: Part I Interconnection (Terminal Attachment).* Ottawa: Minister of Supply and Services, 1981.

Canada. Royal Commission. *Report of the Royal Commission on Corporate Concentration.* Ottawa: Minister of Supply and Services, 1978.

Canadian Satellite Communications. "News Release, Cancom, Telesat form Alliance to Market Hughes VSAT Products." Cancom, Director Communications and Public Relations, November 30, 1995.

Communications Canada. *Telecommunications in Canada, Quick Facts—1987.* Ottawa: Minister of Supply and Services, 1988.

———. *Telecommunications in Canada: An Overview of the Carriage Industry.* Ottawa: Minister of Supply and Services, 1992.

"The Computer Age: Automating the Boss' Office." *Business Week*, April 7, 1957.

Cooper, Mark, and Gene Kimmelman. *The Digital Divide Confronts the Telecommunications Act of 1996: Economic Reality versus Public Policy*. Washington, D.C.: Consumer Union, 1999.

Crandall, Robert W. "Entry Divestiture and the Continuation of Economic Regulation in the United States Telecommunications Sector." In *Deregulation or Re–regulation? Regulatory Reform in Europe and the United States,* ed. Giandomenico Majone. London: Pinter, 1990.

Cutler, Lloyd N., and D. R. Johnson. "Regulation and the Political Process." *Yale Law Journal*. 84 (June 1975): 1395–1418.

Darling, Allan J., "Letter to Janet Yale, AT&T Canada. Re: Unitel Restructuring— Ownership and Control Review." October 16, 1996.

Demsetz, Harold. "Why Regulate Utilities?" *Journal of Law and Economics* 11 (1968): 55–65.

Department of Communications. *Telecommission Study 1 (b), History of Regulation and Current Regulatory Setting*. Ottawa: Information Canada, 1971a.

———. *Instant World*. Ottawa: Minister of Supply and Services, 1971b.

———. The Honorable Gérard Pelletier. *Proposals for a Communications Policy for Canada. A Position Paper of the Government of Canada*. (Green Paper.) Ottawa: Information Canada, 1973.

———. *Consultative Committee on the Implications of Telecommunications for Canadian Sovereignty (Clyne Committee). Telecommunications and Canada*. Ottawa: Minister of Supply and Services, 1979.

———. *Canadian Telecommunications: An Overview of the Canadian Telecommunications Carriage Industry*. Ottawa: Minister of Supply and Services, 1983a.

———. *An Economic Analysis of Enhanced Telecommunications Services*. Ottawa: Department of Communications, Telecommunications Policy Branch, 1983b.

———. *The Impact of Bypass on the Future Developments of Local Telecommunications Networks*. Ottawa: Department of Communications, Telecommunications Policy Branch, 1984.

———. *Convergence, Competition and Cooperation: Policy and Regulation Affecting Local Telephone and Cable Networks*. Report of the co–chairs of the Local Networks Convergence Committee. Ottawa: Minister of Supply and Services, 1992.

Derthick, Martha, and Paul J. Quirk. *The Politics of Deregulation*. Washington, D.C.: The Brookings Institute, 1985.

Doern, Bruce G."Introduction: The Regulatory Process in Canada." In *The Regulatory Process in Canada.*, ed. Bruce G. Doern. Toronto: Macmillan, 1978.

Doern, Bruce G., Ian Hunter, Don Swartz, and V. Seymour Wilson. "The Structure and Behaviour of Canadian Regulatory Boards and Commissions: Multi-Disciplinary Perspectives." *Canadian Public Administration* 18, no. 2 (1975): 189–215.

Donahue, Charles Jr. "Lawyers, Economists and the Regulated Industries: Thoughts of Professional Roles Inspired by Some Recent Economic Literature." *Michigan Law Review* 70 (1971): 195–220.

Dyson, Esther, George Gilder, George Keyworth, and Alvin Toffler. *CyberSpace and the American Dream: A Magna Carta for the Knowledge Age*. The Progress and Freedom Foundation. Release 1.2, August 22, 1994 (http://www.pff.org/position.html). Consulted: October 10, 1999.

Economic Council of Canada. *Reforming Regulation*. Ottawa: Minister of Supply and Services, 1981.

Eisner, Marc Allen. *Regulatory Politics in Transition*. Baltimore, Md.: John Hopkins University Press, 1993.

Faulhaber, Gerald D. *Telecommunications in Turmoil: Technology and Public Policy*. Cambridge, Mass.: Ballinger, 1987.

Federal Communications Commission. *Investigation of the Telephone Industry in the United States*. Washington, D.C.: U.S. Government Printing Office, 1939. Reprint, New York: Arno Press, 1974.

FCC Week. "Appeals Court Throws Out FCC's Computer III Rules." *NewsNet*, June 11, 1990, no. 1.

Ford, D. A., and Associates. *The Impact of International Competition on the Canadian Telecommunications Industry and Its Users: A Study Jointly Sponsored by the Governments of Alberta, Manitoba, Nova Scotia, New Brunswick, Newfoundland, Ontario, Prince Edward Island, Quebec, Saskatchewan, Canada*. Ottawa: D. A. Ford and Associates, 1986.

———. *The Telecommunications Resale Industry in Canada*. Ottawa: D. A. Ford and Associates, 1992.

Friedman, Milton. *Capitalism and Freedom*. Chicago: University of Chicago Press, 1962.

Gabel, Richard. "The Early Competitive Era in Telephone Communication, 1893–1920." *Law and Contemporary Problems* 34 (Spring 1969): 340–359.

Galbraith, John Kenneth. *Economics in Perspective: A Critical History*. Boston: Houghton Mifflin, 1987.

Gillick, David. "Privatization of Teleglobe Canada." *Telecommunications Policy* 11, no. 2 (1987): 213–215.

Glaeser, Martin. *Public Utilities in American Capitalism*. New York: Macmillan, 1957.

Globerman, Steve, with Diane Carter. *Telecommunications in Canada: An Analysis of Outlook and Trends*. Vancouver, B.C.: The Fraser Institute, 1988.

Grant, Peter S. *Canadian Communications Law and Policy: Statutes, Treaties and Judicial Decisions*. Toronto: The Law Society of Upper Canada, 1988.

Green Paper. *See* Department of Communications 1973.

Harding, Robert S. *Register of the Western Union Telegraph Company Collection 1848–1963*. Washington, D.C.: Smithsonian Institution, Archives Center, National Museum of American History, 1990. (http://www.si.edu/organiz/museums/nmah/lemel/dig/westernunion.html). Consulted: February 19, 1999.

Herring, James M., and Gerald C. Gross. *Telecommunications: Economics and Regulation*. New York: McGraw-Hill, 1936.

Hills, Jill. *Deregulating Telecoms: Competition and Control in the United States, Japan, and Britain*. Westport, Conn.: Quorum Books, 1986.

Horwitz, Robert Britt. *The Irony of Regulatory Reform: The Deregulation of American Telecommunications*. New York: Oxford University Press, 1989.

Huber, Peter. *The Geodesic Network: Report on Competition in the Telephone Industry*. Washington, D.C.: U.S. Department of Justice, 1987.

Huber, Peter W., Michael K. Kellog, and John Thorne. *The Telecommunications Act of 1996, Special Report*. Aspen, Colo.: Aspen, 1996.

Intelligent Network News. "Ameritech Files ONA Access Tariffs: Challengers React Quickly." *NewsNet*, February 1991, no. 6.

Janisch, Hudson. "Policy Making in Regulation: Toward a New Definition of the Status of Independent Regulatory Agencies in Canada." *Osgoode Hall Law Journal* 17 (1979): 46–106.

————. "Winners and Losers: The Challenges Facing Telecommunications Regulation." In *Telecommunications Policy and Regulation: The Impact of Competition and Technological Change*, ed. W. T. Stanbury. Montreal: The Institute for Research on Public Policy, 1986.

Janisch, Hudson, and Richard Schultz. *Exploiting the Information Revolution: Telecommunications Issues and Options for Canada.* Commissioned by the Royal Bank of Canada. Montreal: Royal Bank, 1989.

————. "Who Will Regulate Deregulation? Canada's Battle over Telecommunications Competition." *Pacific Telecommunications Review* (September 1992): 17–24.

Kahn, Alfred E. *The Economics of Regulation: Principles and Institutions.* Vol. 1. New York: Wiley, 1971.

Kellog, Michael K., John Thorne, and Peter W. Huber. *Federal Telecommunications Law.* Boston: Little, Brown, 1992.

Kumar, Krishan. *From Post–industrial to Post–modern Society: New Theories of the Contemporary World.* Cambridge, Mass.: Blackwell, 1995.

Lindgren, April. "Crown Sell–Offs Bring $11 Billion." *The Gazette* (Montreal), August 13, 1997, D3.

Lowi, Theodore. *The End of Liberalism.* 2nd. ed. New York: Norton, 1979.

Meyers, Robert A. *Encyclopedia of Telecommunications.* New York: Academic, 1989.

Mosco, Vincent. "Teaching Telecommunications Policy, Critically." *Canadian Journal of Communications* 11, no. 1 (1985): 51–62.

————. *Preuve présentée par le professeur Vincent Mosco au nom du Syndicat des travailleurs et travailleuses en communication et en électricité du Canada. Avis public Télécom CRTC 1991–73, 1991.*

Nader, Ralph. *Unsafe at Any Speed: The Designed-in Dangers of the American Automobile.* New York: Grossman, 1965.

National Telecommunications and Information Administration. *NTIA Telecom 2000. Charting the Course for a New Century.* Washington, D.C.: U.S. Government Printing Office, 1988.

————. *The NTIA Infrastructure Report: Telecommunications in the Age of Information.* Washington, D.C.: U.S. Department of Commerce, 1991.

Noam, Eli M. "Network Pluralism and Regulatory Pluralism." In *New Directions in Telecommunications Policy*, vol. 1, *Regulatory Policy, Telephony, and Mass Media*, ed. Paula Newberg. Durham, N.C.: Duke University Press, 1989.

Noll, Robert. *Reforming Regulation: An Evaluation of the Ash Council Proposals.* Washington, D.C.: The Brookings Institute, 1971.

Pecar, Joseph A., Roger J. O'Connor, and David A. Garbin. *The McGraw-Hill Telecommunications FactBook.* New York: McGraw-Hill, 1993.

Peschek, Joseph G. *Policy-Planning Organizations: Elite Agendas and America's Rightward Turn.* Philadelphia: Temple University Press, 1987.

Posner, Richard. "Natural Monopoly and Its Regulation." *Stanford Law Review* 21 (1969): 551–552.

Québec, ministère des communications. *Proposition d'un partage des responsabilités en matière de télécommunications entre le gouvernement fédéral et les gouvernements des provinces,* 1990.

Rens, Jean-Guy. *L'empire invisible: histoire des télécommunications au Canada de 1846 à 1956*. Vol. 1. Sainte-Foy, Québec: Presses de l'Université du Québec, 1993a.

————. *L'empire invisible: histoire des télécommunications au Canada de 1956 à nos jours*. Vol. 2. Sainte-Foy, Québec: Presses de l'Université du Québec, 1993b.

Rideout, Vanda. *The Continentalization of Canadian Telecommunication Policy: Reorganizing Consent and Increasing Resistance*. Ottawa: National Library of Canada, 1999. Ph.D. thesis, Carleton University, 1997.

Riga, Andy. "Bell Says Rules Cripple It in Long-Distance-Rate War." *The Gazette* (Montreal), February 5, 1997, F4.

Rogers Communications Inc. *72nd Annual Report*. Toronto: Rogers Communications Inc., 1991.

Rohlfs, J. H. "A Theory of Interdependent Demand for a Communications Service." *Bell Journal of Economics and Management Science* 5 (Spring 1974): 16–37.

Roman, Andrew J. "Regulatory Law and Procedure." In *The Regulatory Process in Canada*, ed. Bruce G. Doern. Toronto: MacMillan, 1978.

Rubin, Sandra. "Rogers Doesn't Want More of Unitel: Survival of Money-Losing Long-Distance Company in Jeopardy." *The Gazette* (Montreal), April 20, 1995, D–1.

Schiller, Dan. *Telematics and Government*. Norwood, N.J.: Ablex, 1982.

Schultz, Richard. "Regulatory Agencies in the Canadian Political System." In *Public Administration in Canada*, 3rd ed., ed. Kenneth Kernaghan. Toronto: Methuen, 1977.

————. "Book Review." *Canadian Public Administration* 21 (1978): 291.

————. "Partners in a Game without Masters: Reconstructing the Telecommunications System." In *Telecommunications Regulation and the Constitution*, ed. W. T. Stanbury. Montreal: The Institute for Research on Public Policy, 1982.

————. *Evidence of Richard Schultz. NB Tel Memoranda of Evidence, In the Matter of a Hearing to Review Issues Relating to Interconnection in the Telecommunications Industry in New Brunswick*. Board of Commissioners of Public Utilities of New Brunswick, April 1984.

Simon, Samuel A. *After Divestiture: What the AT&T Settlement Means for Business and Residential Telephone Service*. New York: Knowledge Industry Publications, 1985.

Stanbury, W. T. "Decision Making in Telecommunications" In *Telecommunications Policy and Regulation: The Impact of Competition and Technological Change*, ed. W. T. Stanbury. Montreal: The Institute for Research on Public Policy, 1986.

Stigler, George. "The Theory of Economic Regulation." *Bell Journal of Economics* 2, no.1 (Spring 1971): 3–21.

Stigler, George, and Claire Friedland. "What Can Regulators Regulate? The Case of Electricity." *Journal of Law and Economics* 5 (1962): 1–16.

Stone, Alan. *How America Got On-Line: Politics, Markets, and the Revolution in Telecommunications*. Armonk, New York: Sharpe, 1997.

Stone, Ellery Wheeler. "Interview, April 24th, 1974," *IEEE Oral Histories*. (http://www.ieee.org/organizations/history_center/oral_histories/transcripts/stone9.html). Consulted: March 23, 1999.

Surtees, Lawerence. *Wire Wars: The Canadian Fight for Competition in Telecommunication*. Scarborough, Ontario: Prentice Hall Canada, 1994.

Telesat Canada. *Annual Report*. Gloucester, Ontario: Telesat, Public Relations, 1990.

————. *Telesat Canada Backgrounder*. (http://www.telesat.ca/tcbg.htm). Consulted: August 13, 1997.

Trebing, Harry M. "Telecommunications Regulation—The Continuing Dilemma." In *Public Utility Regulation: The Economic and Social Control of Industry,* eds. Kenneth Nowotny et al. Boston: Kluwer Academic, 1989.

Tunstall, Jeremy. *Communications Deregulation: The Unleashing of America's Communications Industry.* New York: Basil Blackwell, 1986.

Tydeman, John, Hubert Lipinski, Richard Adler, Michael Nyhan, and Laurence Zwimpfer. *Teletext and Videotext in the United States.* New York: McGraw-Hill, 1982.

United States. The President's Advisory Council on Executive Organization, *A New Regulatory Framework: Report on Selected Independent Regulatory Agencies* ["Ash Council Report"] Washington, D.C.: U.S. Government Printing Office, 1971.

U.S. Department of Justice. *Report and Recommendations of the United States Concerning the Line of Restrictions Imposed on the Bell Operating Companies by the Modification of Final Judgment at 6-7, United States v. Western Elec. Co.,* 673 F. Supp. 525 (D.D.C. 1987).

Unitel Communications Inc. "A Profile of Unitel." *Application to Provide Long Distance Service before the Canadian Radio-television and Telecommunications Commission,* 1990: 1–32.

Wadlow, R. Clark, and Richard E. Wiley. *Telecommunications Policy and Regulation 1992.* New York: Practising Law Institute, 1992.

Wilson, James Q. "The Dead Hand of Regulation." *The Public Interest* 25 (Fall 1971): 39–58.

Worthy, Patricia M. *On Competition in the Information Services Industry. Testimony of the Honorable Patricia M. Worthy on Behalf of the National Association of Regulatory Utility Commissioners.* April 20, 1988. United States House of Representatives Committee on Energy and Commerce Subcommittee on Telecommunications and Finance, 1988.

GOVERNMENT DOCUMENTS

Canada

Canadian Radio-television and Telecommunications Commission (CRTC)

Telecom Decision. CRTC 77-10. *Telesat Canada Proposed Agreement with Trans-Canada Telephone System.* Ottawa, 24 August 1977.

———. CRTC 77-16. *Challenge Communications Ltd. v. Bell Canada.* Ottawa, 23 December 1977.

———. CRTC 79-11. *CNCP Telecommunications, Interconnection with Bell.* Ottawa, 17 May 1979.

———. CRTC 80-13. *Bell Canada—Interim Requirements Regarding the Attachment of Subscriber-Provided Equipment.* Ottawa, 5 August 1980.

———. CRTC 81-10. *Bell Canada—Voice Messaging Service Trial.* Ottawa, 25 May 1981.

———. CRTC 81-12. *CNCP Telecommunications—TELENEWS Service.* Ottawa, 18 June 1981.

———. CRTC 81-13. *Bell Canada, British Columbia Telephone Company and Telesat Canada: Increases and Decreases in Rates for Services and Facilities Furnished on a*

Canada-Wide Basis by Members of the TransCanada Telephone System, and Related Matters. Ottawa, 7 July 1981.

———. CRTC 81-22. *Bell Canada—Envoy 100 Service*. Ottawa, 4 November 1981.

———. CRTC 81-24. *CNCP Telecommunications: Interconnection with the British Columbia Telephone Company*. Ottawa, 24 November 1981.

———. CRTC 82-14. *Attachment of Subscriber-Provided Terminal Equipment*. Ottawa, 23 November 1982.

———. CRTC 84-18. *Enhanced Services*. Ottawa, 12 July 1984.

———. CRTC 85-10. *Inquiry into Telecommunications Carriers' Costing and Accounting Procedures: Phase III—Costing of Existing Services*. Ottawa, 25 June 1985.

———. CRTC 85-17. *Identification of Enhanced Services*. Ottawa, 13 August 1985.

———. CRTC 85-19. *Interexchange Competition and Related Issues*. Ottawa, 29 August 1985.

———. CRTC 86-17. *Bell Canada—Review of Revenue Requirements for the Years 1985, 1986 and 1987*. Ottawa, 14 October 1986.

———. CRTC 87-2. *Tariff Revisions to Resale and Sharing*. Ottawa, 12 February 1987.

———. CRTC 88-16. *Bell Canada—ALEX Market Trial*. Ottawa, 30 September 1988.

———. CRTC 90-3. *Resale and Sharing of Private Line Services*. Ottawa, 1 March 1990.

———. CRTC 92-12. *Competition in the Provision of Public Long Distance Voice Telephone Services and Related Resale and Sharing Issues*. Ottawa, 12 June 1992.

———. CRTC 94-4. *Bell Canada—Application to Withdraw ALEX Service*. Ottawa, 24 January 1994.

———. CRTC 94-19. *Review of the Regulatory Framework*. Ottawa, 16 September 1994.

———. CRTC 97-8. *Local Competition*. Ottawa, 1 May 1997.

———. CRTC 97-9. *Price Cap Regulation and Related Issues*. Ottawa, 1 May 1997.

———. CRTC 97-15. *Co-location*. Ottawa, 16 June 1997.

———. CRTC 99-11. *Application Concerning Access by Internet Service Providers to Incumbent Cable Carriers' Telecommunications Facilities*. Ottawa, 14 September 1999.

Telecom Order. CRTC 82-116. *CNCP Market Trial of Infotex Service*. 5 March 1982.

———. CRTC 84-57. *Bell Market Trial of iNet 2000*. Ottawa, 10 February 1984.

Telecom Public Notice. CRTC 1983-72. *Enhanced Services*. Ottawa, 15 November 1983.

———. CRTC 1988-22. *Bell Canada—Market Trial of ALEX Electronic Information and Transaction Service*. Ottawa, 30 May 1988.

———. CRTC 92-78. *Review of the Regulatory Framework*. Ottawa, 16 December 1992.

———. CRTC 1994-130. *Call for Comments Concerning Order in Council P.C. 1994-1689*.

———. CRTC 95-36. *Implementation of Regulatory Framework—Local Interconnection and Network Component Unbundling*. Ottawa, 11 July 1995.

———. CRTC 96-8. *Price Cap Regulation and Related Issue*. Ottawa, 12 March 1996.

Courts

The Bell Telephone Co. of Canada v. Harding Communications Ltd. et al. (1979) 1 S.C.R. at 403.

Supreme Court of Canada. *La Régie des Service Publics v. Dionne* (1978) 2 S.C.R.191; 83 D.L.R. (3d) 181.

Supreme Court of Canada. *Alberta Government Telephones [AGT] and Canadian Radio–television and Telecommunications Commission (CRTC) and CNCP Telecommunications (CNCP)* (1989) 98 N.R. 161.

Laws

An Act Respecting the Bell Telephone Company of Canada—S.C., 1967-68, 48 and the *Bell Canada Act*—S.C. 1987, C.19.
An Act Respecting Broadcasting and to Amend Certain Acts in Relation thereto and in Relation to Radiocommunication (Broadcasting Act), Chapter B-9.01, (Broadcasting Act) Assented to 1st February, 1991.
An Act Respecting Telecommunications 40-41-42 Chapter 38 (Telecommunications Act), Assented to 23 June 1993.
Bell Canada Act. Amendment to Chapter 48 of the Statutes of 1967-68, Section 5.
Bell Canada Special Act, S.C. 1880, c.67. {Special Act of Parliament}
Bell Canada Special Act, S.C. 1977-78, c.44, D.5.(3).
Bill C-16, proposed *Telecommunications Act*, first reading November 9, 1978.
British North America Act of 1867. 30 & 31 Victoria, ch. 3 (U.K.).
Broadcasting Act. S.C. 1967-68, c. 25.
Canada Business Corporations Act. R.S., 1985, c. C-44, s. 1; 1994, c. 24, s. 1(F).
Canadian Overseas Telecommunications Corporation Act. S.C. 1949.
Combines Investigation Act, R.S.C. 1970, c. C-23.
Competition Act. R.S., 1985, c. C-34, s. 1; R.S., 1985, c. 19 (2nd Supp.), s. 19.
Constitution Act of 1867. See *British North America Act of 1867.*
National Telecommunications Powers and Procedures Act. (1985) 1987, S.C. 1987, c. 35.
National Transportation Act, R.S.C. 1970, c. N-17.
Railway Act, Chapter 29 Statutes of Canada 1888.
Supreme Court Act. R.S., c. S-19, s.1.
Teleglobe Canada Reorganization and Divestiture Act, S.C. 1987, c.12.
Telesat Canada Act. 1968-69, c.51, s.1.
Telecommunications Act (1993). See *An Act Respecting Telecommunications* 40-41-42 Chapter 38 (Telecommunications Act) Assented to 23 June 1993.

Other

Order in Council P.C. 1977-3152 (3 November 1977)
———. 1981-3456 (8 December 1981)
———. 1994-1689 (1994)

United States

Courts

AT&T Corp. Tele-Communications, Inc.; TCI Cablevision of Oregon Inc., and TCI of Southern Washington v. City of Portland and Multnomah County. United States District Court for the District of Oregon. June 3, 1999.

California v. *FCC*, 905 F.2d 1217 (9th Cir. 1990).

Dartmouth College v. *Woodward* (17 U.S. (4 Wheat.) 518 (1819)).

Hush-A-Phone v. *United States*, 238 F. 2c at 269 (1956).

Hush-A-Phone Corp. v. *United States*, 238 F.2d 266 (D.C. Cir. 1956), 3.6 n19; 10.4 nn.17–19.

In the Matter of the Use of the Carterfone Device in Message Toll Telephone Service. 13 F.C.C., 2d 240, 13 R.R. 2d (1968).

ITT v. *GTE Corp.*, 449 F. Supp. 1158 (D. Ha. 1978), 3.7 nn.14, 16, 17.

Letter from Nathan C. Kingsbury, AT&T, to James C. McReynolds, Attorney General (Dec. 19, 1913) (Kingsbury Commitment).

MCI Telecommunications Corporation, et al. v. *Federal Communications Commission.* 580 F. 2d (D.C. Cir. 1978). An abridged version of the decision can be found in Brotman (1987, 214–220).

Modification of Final Judgement. United States v. *AT&T*, 522 F. (D.D.C. 1982).

Munn v. *Illinois.* 94 U.S. 113 (1877).

Order, U.S. District Court for the District of Columbia, Civil Action No. 82-0192. September 10, 1987, "Opinion."

Order, U.S. District Court for the District of Columbia, Civil Action No. 82-0192. March 7, 1988, 65.

United States v. *Western Electric Co.*, No 17-49 (D. New Jersey, January 14, 1949).

United States v. *Western Electric Co.*, Trade Cas. (Consent Decree) (CCH) ¶68,246 (D. N.J 1956).

United States v. *Western Elec. Co., Inc.*, 900 F.2d 283, 300-305, 309-310 (D.C. Cir. 1990).

United States Supreme Court 1895. *United States* v. *Union Pac. Ry. Co. et al.* No. 334. November 18, 1895.

Federal Communications Commission (FCC)

FCC. 1949. *United States* v. *Western Electric Co.*, No.17-49 (D. New Jersey, January 14, 1949).

FCC. 1955. *Hush-A-Phone* 20 F.C.C. 391 (1955).

FCC. 1959. *In the Matter of Allocation of Frequencies in the Bands Above 890 Mc*, 27 F.C.C. 359 (1959), *recon. denied*, 29 F.C.C. 825 (1960).

FCC. 1968. *In the Matter of Use of the Carterfone Device in Message Toll Telephone Service.*13 F.C.C., 2d 240, 13 R.R. 2d 597 (1968).

FCC. 1970a. *Domestic Communications-Satellite Facilities (DOMSAT I)*, First Report and Order, 22 FCC 2d 86 (1970a).

FCC. 1970b. *Specialized Common Carrier*, 24 F.C.C. 2d (1970b).

FCC. 1971. *Regulatory and Pricing Problems Presented by the Interdependence of Computer and Communications Facilities (Computer I)*, Final Decision and Order, 28 F.C.C. 2d 267, 269 (1971).

FCC. 1972. *Domestic Communications-Satellite Facilities (DOMSAT II)*, Second Report and Order, 35 FCC 2d 844 (1972a).

FCC. 1972. *In the Matter of Establishment of Domestic Communications-Satellite Facilities by Non-Governmental Entities.* 35 F.C.C. 2d 844, 24 R.R. 2d 1942 (1972b).

FCC. 1977. *MCI* v. *FCC*, 561 F.2d to 365 D.C. Cir. (1977). (*Execunet I*).

FCC. 1978. *ITT* v. *GTE Corp.*, 449 F. Supp. 1158 (D. Ha. 1978), 3.7 nn.14, 16, 17.

FCC. 1978. *MCI* v. *FCC*, 580 F.2d 590 D.C. Cir. (1978). (*Execunet II*).

FCC. 1980. *Amendment of 64.702 of the Commission's Rules and Regulations, Second Computer Inquiry (Computer II)*, Final Decision, 77 F.C.C. 2d. (1980).

FCC. 1985. *Amendment of Sections 64.702 of the Commission's Rules and Regulations, Notice of Proposed Rulemaking (Computer III)*, FCC 85-397 (released August 16, 1985).

FCC. 1986. *In the Matters of Amendment of Section 64.702 of the Commission's Rules and Regulations (Third Computer Inquiry); and Policy and Rules Concerning Rates for Competitive Common Carrier Services and Facilities Authorizations Thereof, Communications Protocols Under Section 64.702 of the Commissions Rules and Regulations* F.C.C. 2d. 60 R.R. 2d 603 (1986).

FCC. 1991. *Computer III Remand Proceedings*, Dkt. No. 90-623 (December 20, 1991).

Laws

Administrative Procedures Act. 5 U.S.C. 706. (1946)

Antitrust Procedures and Penalties Act (Tunney Act.) 15 U.S.C. § 16 (e) (1974)

Communications Act of 1934, 47 U.S.C.

Communications Satellite Act of 1962, Pub. L. No. 87-624, § 102, 76 Stat. 419 (1962)

Mann–Elkins Act de 1910, 36 Stat. 539, 544 § 7 (1910)

Sherman Antitrust Act, 26 Stat. 204, ch. 647 (1980), codified in 15 U.S.C. §§ 1 *et seq.*

Telecommunications Act of 1996, Pub. L. No. 104-104, 110 Stat. 56 (1996)

Willis–Graham Act of 1921, 42 Stat. 27 (1921)

Index

About the Author

Kevin G. Wilson, Ph.D., is professor of communications at Télé-université, the distance learning university of the Université du Québec, where he has been responsible for the BA program in communications since 1988. He has published in the areas of telecommunications policy, media policy, privacy, and the social implications of new communications technologies. He can be reached at kwilson@teluq.uquebec.ca